British Aircraft Guns of World War Two

The Official Air Publications for the Lewis, Vickers and Browning machine-guns and the Hispano 20mm cannon

RAF Museum Series: Volume 9

General Editor: John Tanner, Director, RAF Museum, Hendon

LONDON/MELBOURNE: ARMS AND ARMOUR PRESS
NEW YORK: HIPPOCRENE BOOKS, INC.

Published in Great Britain, 1979,
by Arms and Armour Press,
2-6 Hampstead High Street,
London NW3 1QQ,
and in Australia at
4-12 Tattersalls Lane,
Melbourne, Victoria 3000.

ISBN 0-85368-292-5

Published in the United States, 1979,
by Hippocrene Books, Inc.,
171 Madison Avenue, New York,
N.Y. 10016.

ISBN 0-88254-495-0
Library of Congress Catalogue Card Number:
79-89016

 The material contained in this volume originally appeared under the Air Publication references 120 (*Handbook on the Vickers Gun*), 114 (*Handbook on the Lewis Gun*), 1641B (*Vickers .303in Gas Operated Mk. I Gun*), 1641C (*Browning .303in Mark II Gun*), 1641F (*Hispano 20mm Mk. I and II Guns*) and 1641L (*American Machine Guns*). The original Air Publications were produced for official use and internal purposes by the Air Ministry. The quality of both the type and illustrations in the originals used in the production of this volume fall short of the high standards normally expected in a modern book; they have, of necessity, been reproduced to complete this facsimile edition.

Jacket illustration: The four .303in Browning machine-guns in the power-operated turret of the Boulton Paul Defiant. Aircraft currently on display in the Battle of Britain Museum, Hendon. (Photograph courtesy RAF Museum, Hendon)

Printed by Kingprint Limited, Richmond, Surrey.
Bound by Kemp Hall Bindery, Oxford.

Contents

Foreword	6
1. The Vickers Gun	7
2. The Lewis Gun	25
3. Vickers .303in Gas Operated Mk. I Gun	37
4. Browning .303in Mark II Gun	79
5. Hispano 20mm Mk. I and II Guns	121
6. Browning .50in Guns	181
Amendments	207

Foreword

by J. M. Bruce
Keeper of Aircraft and Aviation Records
Royal Air Force Museum, Hendon

"When my brother and I built and flew the first man-carrying flying machine, we thought that we were introducing into the world an invention which would make further wars practically impossible."—Orville Wright, 1917.

Unfortunately for Orville, the machine-gun had been around for quite some time before the birth of the aeroplane, and it was not long before someone realised the potential of mounting such a weapon on an aircraft. In France, Gabriel Voisin actually mounted a large quick-firing gun on one of his biplanes and displayed it in the Paris Aero Salon of 1910. During that same year, August Euler in Germany patented the idea of installing a fixed machine-gun to fire forward from a pusher aeroplane.

In the years immediately preceding the First World War, the Royal Aircraft Factory and Royal Flying Corps experimented with installations of Maxim, Lewis and Rexer machine-guns, but the RFC and RNAS entered the war with very few guns of any kind and even fewer satisfactory mountings for them. Perhaps fortunately, all the other combatants were in a similar situation.

The combat demands soon imposed by the 1914-18 war led to the establishment of the Lewis and Vickers guns as standard aircraft weapons, not only in the British flying services but in French, Belgian, Italian and American service also. Various refinements and modifications were introduced on both guns, but by the Armistice they remained fundamentally unchanged.

Weapon philosophy changed surprisingly little during the brief inter-war years, British fighters continuing to be armed with rifle-calibre machine-guns. The standard weapons continued to be the Lewis and a refined and lightened development of the Vickers: early investigations into the armour-piercing effectiveness of .5in guns were, remarkably, abandoned. Even more remarkably, they were not revived when it was realised that greater firepower was needed to destroy the more substantial airframes that evolved. The RAF's response to this need was to fit increased numbers of guns in fighters, hence the four-gun Gladiator, succeeded by the eight-gun Hurricane and Spitfire.

For such installations the American Browning gun was adopted in 1934. Also at that time, the RAF sought a replacement for the Lewis as a free-mounted gun. Choice fell on the Vickers Gas Operated Gun, a magazine-fed weapon that was developed from the Vickers-Berthier gun of 1930. It bore a superficial resemblance to the Lewis but, although a good gun in many ways, it was unsuited to the turret installations that were developed for multi-seat aircraft and had to give way to the Browning.

It had been realised in the 1930s that rifle-calibre guns would be less than wholly effective against aircraft fitted with armour, and the RAF sought a heavier weapon. In 1937 two 20mm Hispano guns were acquired for experiments, and the gun was soon adopted for future RAF use. Much development had to be undertaken and many disappointments suffered before the Hispano gun became a truly efficient weapon, but by the summer of 1941 RAF fighter pilots were registering spectacular successes with the Hispano gun. Refinement and development produced the Mark V gun and the belt-feed mechanism.

This volume comprises reprints of the technical descriptions of these guns contained in RAF handbooks. They may reflect little of the ingenuity and skill that lay behind the immense effort of technical development expended on the weapons, yet they are a prosaic commentary on man's seemingly inexhaustible talent for producing ever more means of his own destruction, the same talent that prompted Orville Wright's expression of despair.

1

The Vickers Gun

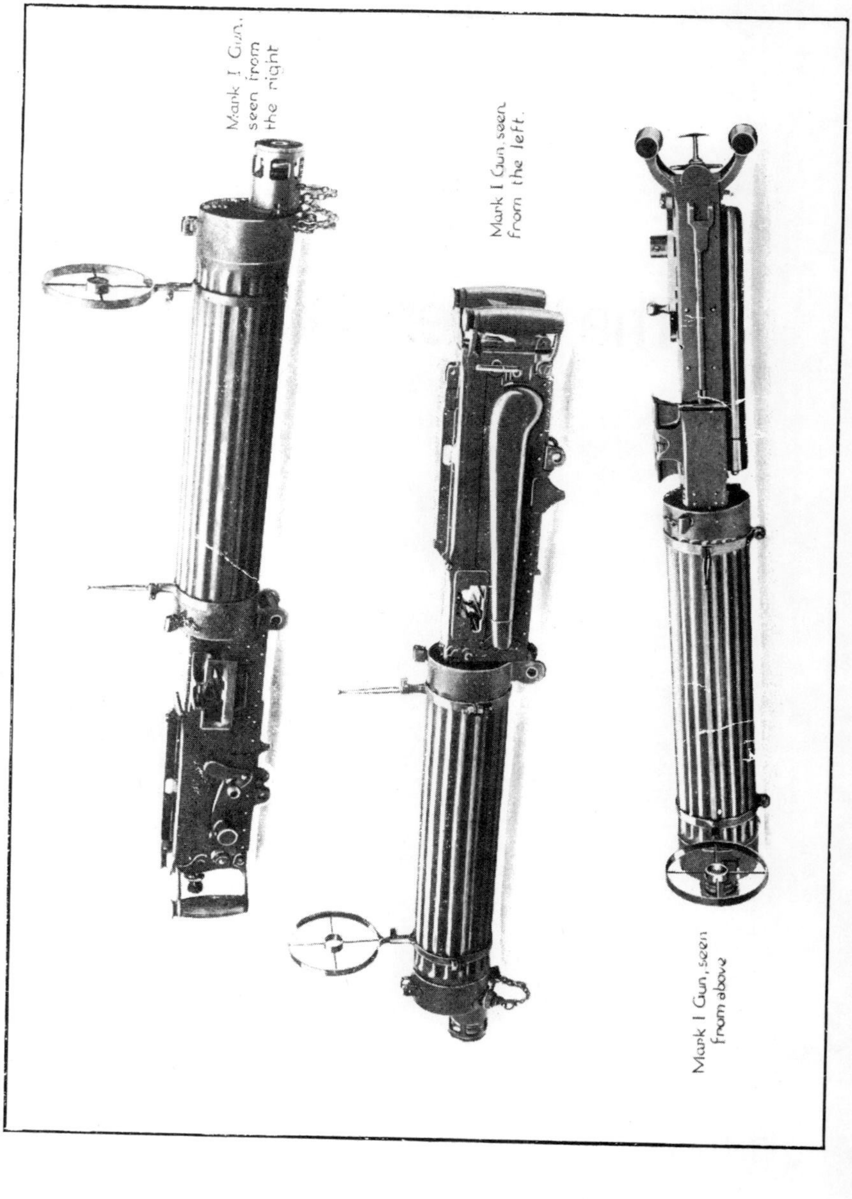
Mark I Gun, seen from the right
Mark I Gun, seen from the left.
Mark I Gun, seen from above

Plate 1.

HANDBOOK ON THE VICKERS GUN.

I. GENERAL DESCRIPTION.

Weight :

Ground gun (Mark I.) **about 28½lbs. with casing empty.**
about 38½lbs. with water in casing.

Aerial gun (Mark ~~II~~ I*.) about 28½ lbs.

Note.—Guns manufactured during the war had a number of refinements omitted, and are about 4lbs. heavier.

Method of operation :

The gun is worked automatically by two forces :

1. The action of the gas generated by the explosion of the charge ;
2. The fusee spring.

Cooling system :

The ground gun is cooled by water which is placed in the barrel casing, the barrel itself being constantly surrounded by the cooling medium.

The aerial gun is cooled by air, the front and sides of the barrel casing being cut away to allow a stream of air to be driven along the exterior of the barrel by reason of the rapid flight of the aeroplane.

General Description of Parts.

The gun is divided into two portions :—

(1) The non-recoiling portions ;
(2) The recoiling portions.

Non-recoiling Portions.

These consist of :—

(*a*) The *Barrel Casing* (including the gland and the outer casing of the muzzle attachment) ;
(*b*) The *Breech Casing.*

(*a*) *Barrel Casing.*

This is made of steel, and has longitudinal corrugations for the purpose of strengthening it.

(i.) *Exterior.*—On the front end of the casing is fitted the *Muzzle Attachment* which consists of two portions, the *Muzzle Attachment Outer Casing* (including the *Front Cone*) and the *Gland.*

The muzzle attachment outer casing is a hollow cylinder threaded internally at its front end to receive the front cone. Behind this portion are six vent holes for the escape of gas. Immediately behind is a threaded hole to take the stud to which is attached the chain carrying the split pin by which the muzzle attachment outer casing and the gland are locked together. Internally, at the rear end, are three rectangular projecting portions which form an interrupted flange for connection with the gland, one of these projections having a small stop. At the centre of each of these projections the outer casing is bored to accommodate the split pin mentioned above. The front cone (Mark II. pattern) is of very substantial construction in order to make it bullet-proof. It is also conical in front so that bullets striking it may be deflected. At the rear of this portion are four serrations on its outer circumference for fixing and removing it. Behind these the diameter is slightly smaller and is threaded to screw into the muzzle attachment outer casing. Behind this the diameter is again slightly smaller and plain. This portion accommodates the rim of the disc. The rear of the cone is bevelled outwards, and then coned inwards, these surfaces being protected by a thin *Disc* of similar shape, so that the cone itself is not eroded by the gases. The cone is bored longitudinally to allow the bullets to pass through it. The gland is circular and hollow. At its front it has three exterior projections which form an interrupted flange corresponding to that on the muzzle attachment outer casing. Beyond these are three holes to take the split pin which locks the gland and the muzzle attachment outer casing together. Immediately behind, on a projecting rim, are four serrations for fixing and removing the part. At the centre of the rear face is a circular projection bored to accommodate the barrel, and threaded externally to screw into the front end of the barrel casing.

At the top of the front face of the barrel casing is a hole to accommodate the screwed head of the steam tube, beside it being a small threaded hole which takes the keeper screw.

For the first two inches of its length, the barrel casing is plain. On the under side of this portion are two holes. **The one on the right is the** *Emptying Hole,* **which is threaded internally to take the** *Screwed Plug* **that closes it. The hole on the left communicates with the** *Steam Escape Tube,* **and is continued into a projection closed by a** *Cork Plug.* **Both these plugs are secured by** *Safety Chains.* At the rear end of the corrugations the barrel casing has another plain surface, **at the top of which is the** *Filling Hole,* **threaded internally to take the** *Screwed Plug* **by which it is closed. (NOTE : This hole is placed slightly to the right in order to prevent the casing from being completely filled, as this might lead to damage during frosty weather, and also to prevent the steam tube from being fouled by the funnel while the casing is being filled).** On the under side of the rear plain portion is a *Bracket* by which the gun is attached to the mounting. Beyond the circular part, the barrel casing is continued at the bottom and towards the rear of the gun. The front end of this continuation forms a *Sleeve* which acts as a bearing for the barrel. Behind this it takes the form of a *Bottom Plate.* This is cut away at the rear end, and strengthened at its edge to serve as a *Seating for Ejection*, while the small portions remaining on each side act as bearing surfaces for the front end of the side plates.

In the aerial gun the barrel casing is cut away at the front, and *Louvres* are made at the rear end of the corrugated portion.

(ii.) *Interior*.—The inside of the barrel casing is tinned, to prevent rust. At the front end is the *Barrel Guide*. This consists of a vertical portion, bored centrally to admit the barrel, and a curved sloping part below to guide the barrel into the hole during the process of assembling. Between the front of the barrel guide and the rear projection on the gland is a small chamber which accommodates the asbestos packing wound on the barrel. Extending backwards to the rear end of the circular portion of the barrel casing is a brass trough which leads to the sloping part of the barrel guide and prevents the tinned surface of the casing being scratched off while the barrel is being assembled.

The *Steam Tube* is made of brass and consists of two tubes. The longer and inner tube runs the whole length of the circular part of the barrel casing. Its rear end is conical and fits into a thimble fixed on the inside of the breech end of the barrel casing. At its front end is a threaded head which screws into the muzzle end of the casing. Near each end of the tube is an entrance hole for steam. In the screwed head is a third hole through which the steam passes to the steam escape tube, the latter being fixed to the inside of the barrel casing. The screwed head is positioned by a keeper screw which ensures that the steam entrance holes are at the top of the tube, and that the third hole is opposite to the steam escape tube. The shorter and outer tube is free to move along the inner tube and is of such length that the steam entrance holes are never both covered at the same time. It thus forms a *Sliding Valve*. This device allows the steam to escape, while the water is retained.

(*b*) *Breech Casing*.

(i.) *Right and Left Outside Plates*.—These extend from the circular portion of the barrel casing to the rear of the gun. At the front end they are riveted, for about six inches, to the extension of the barrel casing. At the top there is in each case a hole for the front cover catch. Just behind, both plates are cut away at the top to accommodate the feed block. Immediately behind this cut-away portion there is on each plate a projection, bored to take the joint cover pin. Along the upper edges both plates carry flanges which fit into grooves on the rear cover, thus ensuring stability. At the rear end the plates have rectangular slots in which the crank bearings move, these slots being partially closed by the *Right Slide* and the *Left Slide*. Above the slotted portion each plate has a hole to accommodate the T-fixing pin. At the bottom in each case is another hole which takes the rear crosspiece axis pin, the corners of the plates being here rounded to give clearance for the movement of the rear crosspiece. At the rear of the plates are two projections which fit into slots in the rear crosspiece. On the under side of each plate is a lug bored centrally, these lugs forming a *Bracket* by which the gun is attached to the mounting.

On the inside of each plate is a cam, by means of which the movement of the extractor is partially controlled. Near the rear end of these cams, and on the top sloping surface, a step is cut, so that the lock may be hindered from going forward if the recoil action is insufficient. These steps also enable the lock to be "hung."

(ii.) The details given above are common to both plates. Peculiar features possessed by either are as follows:—Right outside plate (exterior). Rather more than halfway along there is the *Check Lever Bracket*, which is a cut-away rectangular plate riveted in position. Fastened to this bracket is a stud on which the *Check Lever* is pivoted, and to which it is secured by a small keeper pin. The check lever is cut away to receive the end of the crank handle when the gun is in the firing position. The right slide carries the *Roller*, this being held in position by a *Collar* and *Split Pin*.

Left outside plate (exterior). Near the front end and at the top is a stop which limits the movement of the lever of the front cover catch, the plunger of the latter being accommodated by a groove in the stop. Immediately behind and below are two studs for holding the front end of the fusee spring box. Longitudinally the plate is milled out (ribs being left for strengthening purposes). **Rather more than half-way along, and at the bottom, is the** *Elevating Stop*, **which protects the fusee spring box from damage when the gun is on a tripod mounting.** The left slide carries a stud which forms the back anchorage for the fusee spring box.

(iii.) *Front Cover.* This closes the top of the recess which accommodates the feed block. At its front end, on the lower side, is riveted a plate carrying the *Hooks* by which it is secured. Behind this its lower surface is milled out almost to the rear, a rim being left to fit inside the flanges on the feed block. For a short distance, at the front, this rim is milled externally to give a grip when the cover is raised. At the rear end, on the underside, is a projection forming the *Extractor Stop*. Where the front cover is hinged to the rear cover, it has an enlarged portion, cut away in the middle, and bored to accommodate the *Joint Cover Pin*.

The *Front Cover Catch* extends from the right to the left outer side plates, and comprises a cylinder having two flats cut into it. On the left is the *Front Cover Catch Lever*, which is in the shape of a handle, the horizontal arm being hollowed out to contain the *Front Cover Catch Plunger*, *Spring* and *Plug*. When the lever is turned up, the front cover hooks are in line with the flats on the front cover catch, and the cover may be raised. When the lever is lowered, the front cover hooks are retained by the circular portion of the catch. The outer end of the lever is milled, to form a finger-grip. The catch is bored near the left end to receive a split pin which prevents its withdrawal from the gun.

(iv.) *Rear Cover.* At the front end of the rear cover is an enlarged portion cut away on each side and bored to admit the joint cover pin. On the under side are grooves, on the extreme right and left, to accommodate the flanges on the outer side plates. The under side is milled out into three channels, **the centre one of which accommodates the trigger bar,** the ribs on either side forming guides for the top of the lock. In the side channels are riveted the *Ramps,* which are, roughly speaking, triangular in shape, the longest side being curved. On the sides of the centre channel are three pairs of small projections **by which the trigger bar is retained in position. The** *Trigger Bar* **has on either side, from front to rear, a rim which is held by these projections. At its front end the trigger bar is slotted for nearly half its length. Near the rear end it has on the right a projection which serves as a seating for the** *Trigger Bar Spring,* **the spring itself being spiral and contained in a tunnelled projection at the rear of the right channel. The extreme rear of the trigger bar has a small portion turned downwards for the action of the trigger bar lever.**

At the rear of the cover are two cut-away projections bored to accommodate the *Rear Cover Lock Axis Pin.* The *Rear Cover Lock,* which pivots on this pin, is right-angled in shape, and consists of a small semi-circular plate with three projections. Two of these are in the form of hooks which engage in recesses in the rear crosspiece. The third is a lug on which the rear cover lock spring acts. The curved plate extends behind the breech casing and serves as a thumb-piece for lifting the cover. The movement of the rear cover lock is limited by two small lugs, one on each side, which engage in recesses in the projections at the rear of the cover itself. **On the under side of the lock is a recess which gives clearance for the movement of the fore part of the safety catch.** The rear cover lock is normally held down by the action of the *Rear Cover Lock Spring,* which is flat and lies on the left channel on the under side of the cover. The front end of this spring is slotted to fit under a T-shaped stud which serves it as a seating. The rear end of the spring presses on the lug on the cover lock as already described.

(v.) *Rear Crosspiece.*—This comprises a plate which closes the breech end of the gun. On its inner side, to right and left, are ribs forming two grooves which accommodate the rear edges of the outer side plates. At two points on each side these ribs have enlarged portions bored to receive respectively the T-fixing pin and the *Rear Crosspiece Axis Pin.* On the inner side of the upper pair of these projections are small curved seatings for the horizontal arms of the trigger bar lever. At the bottom, joining the lower pair of projections, is a horizontal plate which, when in its normal position, forms a part of the bottom plate of the breech casing. At the top, on the left, the inner side of the plate is bevelled to give clearance for the movement of the rear cover lock spring. In the middle, near the top, are three holes, **the centre one being circular to admit the safety catch piston and spring,** while those to right and left are rectangular and provide a grip for the hooks of the rear cover lock. **Below, and in the middle, is another rectangular hole which admits the pawl on the firing lever. The rear face of the plate of the crosspiece has two pairs of projections bored to accommodate respectively the safety catch axis pin and the firing lever axis pin.** At the top the plate is bevelled to right and left for the guidance of the hooks on the rear cover lock. **In the middle of the plate is a vertical recess which admits the lug on the safety catch.** On either side of this recess, at the extreme right and left, are rectangular slots which receive the rear projections on the outer side plates. Extending from the corners of the plate are brackets, to which are fitted the *Traversing Handles.** These handles are hollow steel cylinders covered with wood. They form small reservoirs for oil, and are closed by milled plugs which screw into the cylinders and carry camel hair brushes on the under side.

Trigger Bar Lever.—**This is cruciform in shape. At the top it is flattened to fit closely the rear end of the trigger bar. Immediately below the top, on its rear side, is a circular recess to receive the safety catch spring. At the bottom, on the same side, is a horizontal recess for the pawl on the firing lever. The horizontal arms of the trigger bar lever are bored to take the T-fixing pin, which acts as its axis.**

Safety Catch.—**This consists of a central part with extensions to right and left, the ends of these being flattened, enlarged and milled on the under side, to form finger grips. The central part is bored to take the** *Safety Catch Axis Pin,* **and has at the top a projection which limits the amount of upward movement of the safety catch. On the under side two stops are provided which control and limit the movement of the firing lever. The safety catch is normally held in the " safety " position by the** *Safety Catch Piston and Spring,* **the former being a small hollow cylinder with a rectangular head, while the spring is spiral.**

Firing Lever.—**At the top this is slightly cut away to fit the stops of the safety catch. On its inner side it is recessed to accommodate the** *Firing Lever Pawl.* **Towards its lower end it increases in thickness, and at the bottom it is bored to receive the** *Firing Lever Axis Pin.* **At its lower extremity, on the inner side, is a flattened portion which acts as a stop. Extending to the rear, at the top, is the** *Firing Lever Thumbpiece,* **which is T-shaped and has a curved flat plate milled on the outer side. The firing lever pawl is wedge-shaped and bored to accommodate the axis pin on which it is pivoted.**

T-Fixing Pin. This, as the name indicates, is T-shaped. The longer arm is threaded externally at the end to screw into the flange on the right of the rear crosspiece. Near the other end of this arm is a collar which positions it on the flange on the left side of the rear crosspiece. The head of the T-fixing pin is of two different diameters, the smaller of which enables it to be used to push out the split pin from the side levers axis bush when the lock is being stripped. For this purpose the extremity of this arm is slightly hollowed out.

(vi.) *Bottom Plate.*—This consists of three portions, which, together with the rearward extension on the barrel casing, completely close the under side of the gun. The largest part of the bottom plate comprises a fixed portion which is cut away for about half its length in front. **This aperture is closed by a sliding shutter which is accommodated in grooves on the under side of the fixed plate. At the front end of the** *Sliding Shutter* **is a turned-up portion which connects with the extension of the barrel casing. Below, and just behind, is a transverse tunnelled projection containing a plunger on the left and a** *Sliding Shutter Catch* **on the right, these two being separated by a spiral spring. The plunger limits the forward movement of the shutter**

* The aerial gun has only one Traversing Handle, on the right.

by its contact with a raised stop on the left side of the gun. The sliding shutter catch is operated by a milled projection and a milled circular plate, these affording a grip. On its upper side is a small stud, which, under the influence of the spiral spring, engages in recesses on the right of the bottom plate at the limits of the forward and backward movement of the shutter. Behind the sliding shutter the fixed part of the bottom plate is not cut away. The aperture behind the fixed portion of the bottom plate is closed by a small plate on the rear crosspiece.

(vii.) *Feed Block :* EXTERIOR.—The feed block is made of steel or gun-metal, and fits into a recess in the outside plates of the breech casing. It is covered and held down, when in position, by the front cover. At the front is a projection bored vertically to accommodate the axis pin of the top and bottom levers. Ribs are provided for positioning the feed block in the gun. Behind the projection the feed block is milled out at the top and at the bottom, ribs being left for positioning and strengthening. Immediately to the rear of this milled portion the feed block is cut away in the middle, and has transverse grooves to accommodate the slide. On the right, opposite the cut-away portion, the upper surface of the feed block is recessed. At the rear, the upper part is milled out, ribs being left as before. The rear face is cut away to allow the cartridge to be drawn out, the part on the left of this aperture acting as a *Cartridge Stop*. There are also grooves to accommodate the ribs on the face of the extractor. At the top of the right groove, a small projection is left as a guide for the cartridge rim. INTERIOR. The feed block is hollowed out to admit the cartridges, guides being provided at front and rear to ensure that the rounds are brought to exactly the correct position. At the bottom is a rib which acts as a bearing surface for the belt as it is fed in. At the front, on the right, is a bevelled portion which serves as a guide for the noses of the bullets, while towards the left side are two projections which form the *Bullet Stop*. At the rear are two raised portions, above and below, and rounded off on the right side, which serve as guides to the rims of the cartridges. At front and rear, and slightly to the right, are slots through which the bottom pawls project into the feed block.

Bottom Lever.—This is right-angled in shape. The horizontal arm is flat and has on the under side a semi-circular projection which engages in the recess in the prolongation of the left side-plate. The other arm is bored to receive the split pin which secures the part to the top lever. The vertical portion of the lower lever is cylindrical, except at its upper end, where it is hexagonal.

Top Lever.—This is bored at its front end to fit the hexagonal part of the bottom lever. At the rear end, on the under side, is a stud which engages in the slot in the slide. The top and bottom levers are secured together by a split pin having a flat head.

Slide.—This moves transversely in the feed block, under the influence of the stud on the top lever, the stud being accommodated by an elongated hole bored through the slide, slightly to the left. On the under side it is for the most part milled out. On the left are two projections which form the seating for the top pawls spring. On the right are two other projections bored to receive the top pawls axis pin, The right side of the slide has two small recesses to accommodate projections on the top pawls.

Top Pawls.—These are two in number. They are bored to receive the *Top Pawls Axis Pin* on which they pivot. The left extremity in each case is milled, to control the movement of the rounds more effectively. The front pawl has also an enlarged head. At the right extremity of each pawl is a small curved plate milled to form a finger-piece. On each of these finger-pieces is a small projection fitting into the slide.

Top Pawls Spring.—This is flat and U-shaped, and carries on the left a projection for fixing and removing it.

Bottom Pawls.—These are connected with each other by means of a curved milled plate, to which they are riveted, the three portions forming one U-shaped, right-angled part. The pawls are bored to admit the *Bottom Pawls Axis Pin* on which they pivot, this pin having an enlarged head at one end and a split portion at the other.

Bottom Pawls Spring.—This is a double spring which is retained in position by the bottom pawls axis pin.

(viii.) *Fusee, Fusee Chain, Fusee Spring and Fusee Spring Box.*

Fusee.—This consists of a curved portion cut away at the front to a right-angled recess which fits the links of the fusee chain. At the upper end of this part is a hole bored to admit the pin by which the fusee and the fusee chain are fastened together. At the lower end of the fusee, and on the inner side, is a stem with two lugs and two feathers which position it correctly in the crank.

Fusee Chain.—This consists of two links, the forward link being solid, while the other one is hollow. At the front end of the solid link are two small projections which fit into the hooks at the rear of the fusee spring.

Fusee Spring.—This is a strong spiral spring fastened at its front end to a circular brass plug which is bored and threaded internally to receive the adjusting screw. At the rear end of the spring is fastened an attachment which has two hooks projecting to the rear.

Adjusting Screw.—At the front end this is bored to admit the *Vice Pin* by which the tension of the fusee spring is adjusted. The head of the adjusting screw is retained on the outer side of the fusee spring box by the front end of the box. Behind the head of the tension screw are two small projections which fit into recesses in the front of the fusee spring box. To the rear of the head the adjusting screw is threaded externally to connect with the brass seating on the fusee spring.

Fusee Spring Box.—This is an elongated casing provided with two hooks at the front end and one at the rear, by means of which it is attached to the left outer side plate and the left slide. At the front end are two small recesses to accommodate the projections on the rear of the head of the adjusting screw. The upper edge of the box is slightly cut away near the front to clear the feed block. The lower edge is cut away towards the rear to clear the middle rib on the left outer side-plate

Recoiling Portions.

These consist of :--

(a) The *Barrel* (together with the *Muzzle Cup*).

(b) The *Side-plates*.

(c) The *Crank* (including *Crank, Crank Handle*, and *Connecting Rod*).

(d) The *Lock*.

(a) *Barrel.*—This is 28½ inches in length. At the fore end it provides accommodation for the muzzle cup. On the Mark I barrel this takes the form of a groove immediately behind the muzzle, for the clamping-screw, a flat being cut in front of the groove to allow the cup to be assembled. In the Mark II barrel the fore end is threaded to take the muzzle cup. The barrel is tapered towards the front. Near the rear end is a *Cannelure* (or groove) to accommodate asbestos packing. The part of the barrel immediately behind and before the cannelure is reinforced, and the external diameter is therefore greater. The breech end of the barrel forms a square *Barrel Block* from which the two *Barrel Trunnions* project at the sides. At the rear of the barrel are vertical grooves to accommodate the ribs on the face of the extractor. The barrel is bored longitudinally and rifled, having five grooves and lands with a left-handed twist. Above the rear end of the boring is a small cut-away portion which guides the noses of the bullets into the chamber. Inside the reinforced part at the rear, the barrel is shaped to form a *Chamber* into which the cartridge exactly fits. Immediately in front of this chamber is the *Lead*, which forms a funnel to guide the bullet into the rifling.

Muzzle Cup.—This consists of the cupped portion itself and stem for attachment to the barrel. It is bored from front to rear to admit the barrel, and on the outside of the stem are four serrations for fixing and removing it. In the Mark I pattern the stem is split, and bored transversely to accommodate a clamping-screw which lies in the groove on the muzzle end of the barrel. In the Mark II pattern the stem is not split, but is threaded internally in front to screw on to the barrel. Except where the conversion set is used for speeding-up the gun, a .05 steel washer should be placed between the end of the barrel and the Mark II muzzle cup.

(b) *Left and Right Side-Plates.*—For the greater part of their length both plates are milled out on the outer side. At the front end they are bored to receive the barrel trunnions. Near the front end they are fitted with *Side-Plate Springs* which ensure that the horns of the extractor do not drop below the cams on the outer side-plates during the backward movement of the lock, when there are no cartridges on the extractor. At the bottom of each side-plate is a guide, from front to rear, along which the flanges of the lock move. Each of these guides has two interruptions to enable the lock to be lifted out. Near the rear end each plate has a square reinforced portion with a circular boring which acts as a bearing for the crank. Immediately behind, both plates have extensions for the exclusion of grit from the slots in the outer side-plates. At the front end of the left side plate is a prolongation with a recess for the engagement of the stud on the lower lever of the feed block.

(c) *Crank.*—This is cut away on both the upper and the under side (to allow for the movement of the sear and the connecting-rod respectively, during the action of the gun). At the front it is bored transversely to admit the crank pin. At the rear of the crank are circular projections to right and left, which are carried in bearings in the side-plates. The projection on the left is bored to accommodate the fusee, the feathers and lugs on the latter being received in featherways and a groove, so ensuring correct fit and driving action. The projection on the right carries the crank handle, for which purpose it is hexagonal in shape at the extremity. The handle is secured in position by a screwed fixing-pin.

Connecting-Rod.—This pivots freely upon the crank pin. At the front it is circular and fitted with three projections which secure its connection with the lock. About half-way along is an exterior thread which accommodates the *Adjusting Nut.* By this means—washers being inserted below the adjusting nut, if necessary—the position of the connecting-rod in relation to the lock is regulated, this in turn regulating the distance between the extractor and the barrel. The rear part of the connecting-rod is rounded and milled on its upper surface to provide a finger-grip.

Crank Handle.—This forms, roughly-speaking, a flat curve. At the front end it is fitted with a knob set horizontally and at right angles. Rather more than half-way along there is on the right side a boss bored hexagonally to fit the right extremity of the crank. On the upper side, at this point, is a shallow depression. The rear part of the crank handle is in the form of a curved, tapered *Tail.*

(d) *Lock.*—This comprises several parts, as follows :—

Casing.—This is cut out of solid metal. The exterior is for the most part slightly milled out, raised portions being left to form ribs along the top and front edges, flanges at the bottom, stops at the front bottom corners, and seatings for the axis pins of the tumbler and trigger and the side levers axis bush. On the face of the casing there is at the top a projection which forms an *Extractor Stop*, and which, in conjunction with the stop on the under side of the front cover, limits the upward movement of the extractor. Near the bottom the face is bored to admit the firing pin. The sides of the casing are bored to accommodate the axis bush of the side levers, the trigger axis pin and the tumbler axis pin. On the left side there are also recesses for the head of the trigger axis pin and the feather on the tumbler axis pin. Near the top rear corner the casing is bored to receive a pin and distance piece by which it is strengthened and the sides correctly positioned. At the bottom are two *Flanges* which ride on the guides of the side-plates during the movement of the lock. In the upper flanges are rectangular slots which accommodate the projections on the side levers. The lower flanges have interruptions agreeing with those in the guides on the side-plates, in order that the lock may be lifted out of the gun. At the front end the bottom of the lock is bored to give clearance to the movement of the head of the sear.

The interior of the casing is entirely hollowed out. Behind the top of the face is riveted a piece of metal which acts as a guide for the lock-spring when the lock is being assembled, and which also forms a seating for the spring. Near the bottom are horizontal ribs from front to rear which act as guideways for the firing-pin. At the front lower corner is a pin which serves as an axis for the sear.

Side Levers.—These are two in number, and, together with the *Side Lever Head*, are cut out of the solid, the whole forming one Y-shaped piece. The side levers are bored and recessed at the front to accommodate the side levers axis bush. On the inner side, round this hole, there is in each case a reinforced portion which serves as a bearing and a distance piece. Immediately behind, and on the under side of the levers, are small projections which fit into slots on the lock casing. The side levers have arms projecting upwards and backwards from the front. At the top these arms are rounded off towards the rear, each curved portion being interrupted by a *Step* which engages with the bent on the extractor levers. The side lever head forms a cylindrical stem by which the lock is joined with the connecting-rod. It is bored longitudinally to admit the stem of that part, and at the front end it has on the inside three projections which engage with the three on the connecting-rod when the lock is in its normal position in the gun.

Extractor Levers.—These are two in number, and take the form of plates of irregular shape. When in position they pivot on the tumbler axis pin and lie close to the outside of the lock casing, the front portion being slightly set out so as to clear the ribs on the edges of the face of the casing. The upper part of the front ends is shaped so as to lift the extractor easily and continuously during its upward movement, while on the under side, and slightly to the rear, is a straight portion which bears on the stop on the side of the lock casing, so limiting the downward movement of the extractor. At the rear end, on the outer side is the *Bent*, which is acted upon by the side lever. At the rear lower corner the extractor levers are bored to receive an axis pin. The right lever is slightly chamfered to clear the small auxiliary cartridge rim guide in the right groove of the breech end of the barrel.

Trigger.—Above the *Nose* (or lower end) of this is a circular portion which serves as a bearing, and which is bored to receive the *Trigger Axis Pin*.

Tumbler.—This is curved in shape. Its *Head* is enlarged and flattened for engagement in the recess in the firing-pin. The central part is enlarged and has a circular rim bored to accommodate the *Tumbler Axis Pin*. Immediately opposite is a *Bent* in which the nose of the trigger engages under the influence of the short arm of the lock-spring. The *Tail* curves slightly downwards to the rear.

Firing-Pin.—This is roughly rectangular. On the upper side there is a projection at the front, against which the long arm of the lock-spring bears. Protruding from the front is the tapered portion which explodes the cartridge. Behind this the firing-pin is cut away so that the side levers axis bush (which runs through it) may not hinder its forward and backward movement. Towards the rear, on the upper side, is a recess in which the head of the tumbler engages. On the under side, halfway along, is a bent—bevelled towards the rear—which engages with a corresponding bent on the sear.

Sear.—This has an enlarged head recessed to form a *Claw* which accommodates the pin inside the front lower corner of the lock casing, by which means the sear is correctly positioned. Behind, and on the under side, is a small slot which forms a seating for the flat *Sear Spring.* On the upper side is a *Bent* which engages with that on the firing-pin. The *Tail* tapers slightly and is turned down at the end.

Lock-Spring.—This is U-shaped. It has two arms, curved and of unequal length, the shorter of which acts on the trigger while the longer one presses on the projection at the front of the firing-pin. The ends are both slightly rounded.

Side Levers Axis Bush.—This is a small hollow cylinder with a rim at one end, this rim being accommodated in a recess in the side levers. Through the bush is passed the *Side Levers Axis Bush Split Pin,* which has a flat head, while its other ends are slightly thicker.

Tumbler Axis Pin.— This has near one end a collar bearing a feather, both being accommodated in the left side of the lock casing.

Trigger Axis Pin.—This has a flat head received in a recess in the left side of the lock casing.

Extractor.—On the front of this, at either side, are ribs, the portion immediately behind being cut away to form *Cartridge Grooves.* At the top the ribs curve forward slightly, the grooves thus being wider to ensure that the rim of the round is gripped. In the face of the extractor is a vertical slot cut to accommodate the gib. Below this is a hole to allow the front end of the tapered portion of the firing-pin to pass through. At the side of the extractor, near the top, are the projecting *Horns,* which support it on the cams during the backward movement of the lock. At the front these horns are fined down to an edge so as to avoid the risk of jamming against the edge of the cams. Below the horns the extractor is cut away on either side, *Shoulders* being left for the extractor levers to bear upon when lifting it. Below these recesses are the *Grooves* by which the side-plate springs grip the extractor. Behind its face the extractor is for the most part cut away. At the top is the recess in which the extractor stop on the face of the lock is accommodated, underneath being the stop which engages with it. Immediately below is the seating for the gib and gib spring, alongside being the grooves provided for the gib spring cover. Near the bottom is the hole for the firing-pin, this aperture being coned to fit the shape of the firing-pin, and also slightly recessed at the top. From the top to the bottom of the extractor are grooves which accommodate the ribs at the side of the face of the lock, and which serve as guides for the extractor during its movement.

Gib.—This has two projecting portions on the front, the upper projection being smaller and curved at the top, while the lower one is bevelled. The gib is retained by flanges at the side, these lying inside the extractor, while the projections protrude through the face of it. The gib is slightly curved at the back to accommodate the flat *Gib*

Spring by which it is held forward. The *Gib Spring Cover* is a thin plate accommodated by grooves and serving as a seating for the spring.

Conversion Set for Speeding-up Vickers Gun.

In order to make Vickers Guns more certain of functioning at high altitudes, and also to increase the rate of fire, Conversion Sets are used. As, however, this speeding-up causes considerable extra strain to various parts of the gun, the conversion set should only be fitted to guns properly adapted to its use. This adaptation involves two features :—

1. The insertion of special parts ;
2. The strengthening or modification of existing parts.

The special parts comprise the following :—

Muzzle Cylinder.—This is a short cylinder which serves as a lining to the muzzle attachment. Round the sides are bored small holes for the escape of gases generated by the explosion of the charge. One end is entirely open, while the other is closed except for a hole to allow for the passage of the bullets. The part around this hole is reinforced. On the outside, at this end, is a threaded portion by which the cylinder is screwed into the fore end of the muzzle attachment outer casing. Beyond is a flange with four serrations (identical with those in the front cone) by which it is screwed on or removed.

Buffer Spring.—This is a strong spiral spring, made in $3\frac{3}{4}$ to $4\frac{1}{4}$ flattened coils and placed over the barrel behind the muzzle cup.

Brass Split Pin.—This is placed through the roller collar on the right slide, and is rendered necessary by the fact that the standard split pin is liable to work out through vibration.

When the Mark I Gun is used, the parts needing modification are :—

Left Side Plate.—Wear takes place with the ordinary pattern after about 4,000 rounds, causing insufficient movement of the feed block, slight cross feeds, and eventual breakage of the right side extractor horn. To avoid these results, the left sideplate must be hardened at the rear end of the recess in the prolongation.

Front Cover Hooks.—It has been found that the old pattern hooks are liable to turn back and allow the front cover to fly open. It is consequently necessary to strengthen the hooks by brazing a small stiffening piece to the front of them.

Rear Cover Lock Spring.—It has been found that the extra speed of the gun may cause the rear cover to fly open. With the conversion set, therefore, a stronger spring must be used, although—as a purely temporary measure—it is possible to strengthen the standard spring by giving it a little more " set."

Fusee Spring.—The ordinary pattern of fusee spring is made from hard drawn wire. For use in the speeded-up gun, springs that have been hardened and tempered are necessary.

Crank Handle.—In order to eliminate the somewhat harsh blow at the close of the recoil action, a portion of the crank handle must be removed, in accordance with the diagram on p. 20. The gun then functions more smoothly, and the number of breakages is reduced.

Mark ~~II~~ I* (Aerial) Gun.

These guns are prepared for aerial purposes only, and possess the following modifications—made before issue :—

All modifications (as shown above) rendered necessary by the use of the conversion set.

The foresight is removed from the barrel casing.

The steam tube and its keeper screw are absent, together with the plugs and safety chains of the emptying hole, the steam escape hole, and the filling hole.

The front cover has four elongated holes for the fitting of the C.C. Gear " C " Type trigger motor.

The rear cover has the tangent sight bridge removed, and is slotted to accommodate the end of the trigger motor push rod of the " C " Type C.C. Gear.

There is no fusee spring box, the fusee spring adjusting screw being positioned by a skeleton anchorage.

The left traversing handle is removed from the rear crosspiece.

The rear crosspiece is bored and tapped to receive the rounds-counting mechanism.

The following parts of the ordinary firing mechanism are absent : Safety catch (together with its piston and spring), firing lever, trigger bar, and trigger bar lever.

The left and right outer side plates are bored to accommodate the electrical gun heaters.

The top of the lock is hollowed out slightly at the front end to accommodate the trigger motor push rod of the " C " Type C.C. Gear.

The extension at the rear of the right side-plate is bored to accommodate the right-angle bracket of the rounds-counting mechanism.

The sliding shutter and catch are removed.

The elevating stop is removed.

The cams are slightly cut away at the front in a diagonal fashion from top to bottom.

The breech end of the barrel is slightly cut away at the top.

Method of Instruction.

The teaching of this branch of the work calls for a considerable effort of memory. It should, therefore, be taken in short sections, each of which must be thoroughly understood before the next is attempted.

The general description should be brief, the main idea being to call the attention of the pupils to the peculiarities of the various parts, with a view to their subsequently understanding the mechanism of the gun.

It has been necessary in a few places to introduce parts not yet taught, for the clearer explanation of the particular portion under consideration. On the rare occasions where this is necessary, the instructor should refer to the unknown part without giving its name. The stage at which each new name should be introduced is indicated by italics.

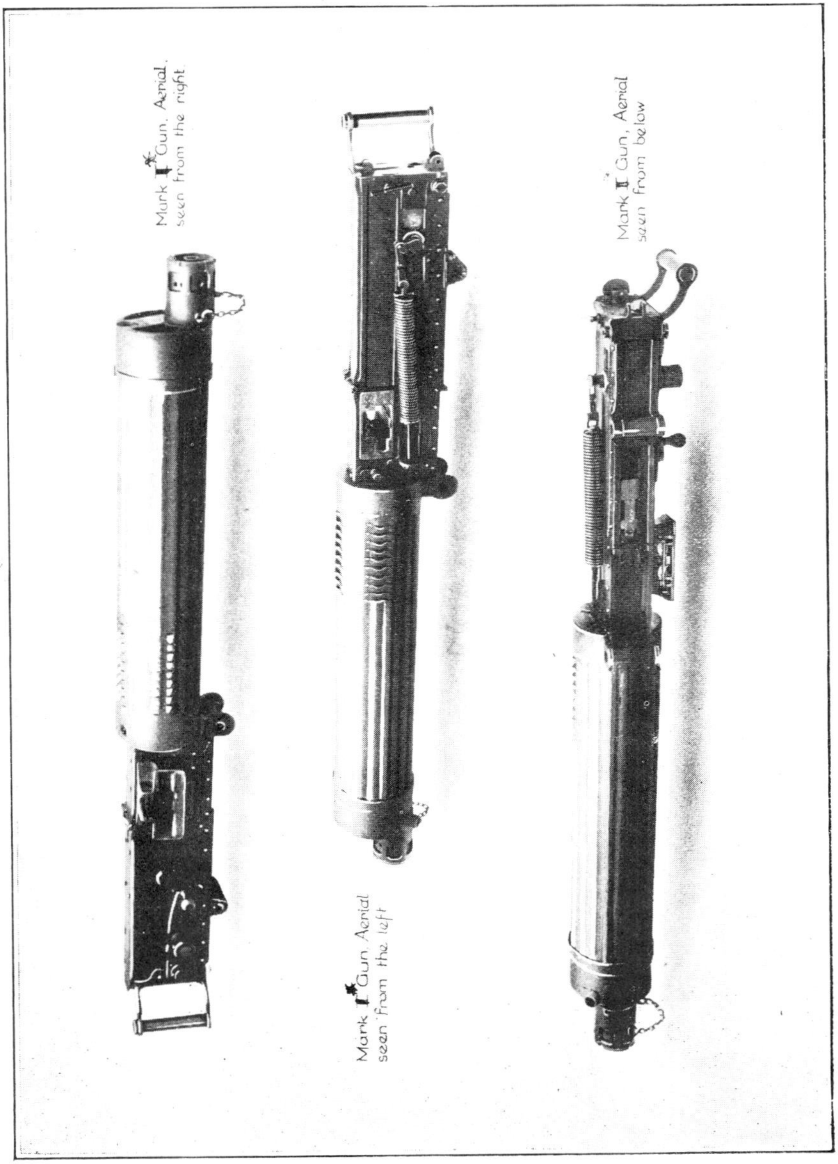

Mark I* Gun, Aerial, seen from the right.

Mark I* Gun, Aerial seen from the left

Mark II* Gun, Aerial seen from below

2

The Lewis Gun

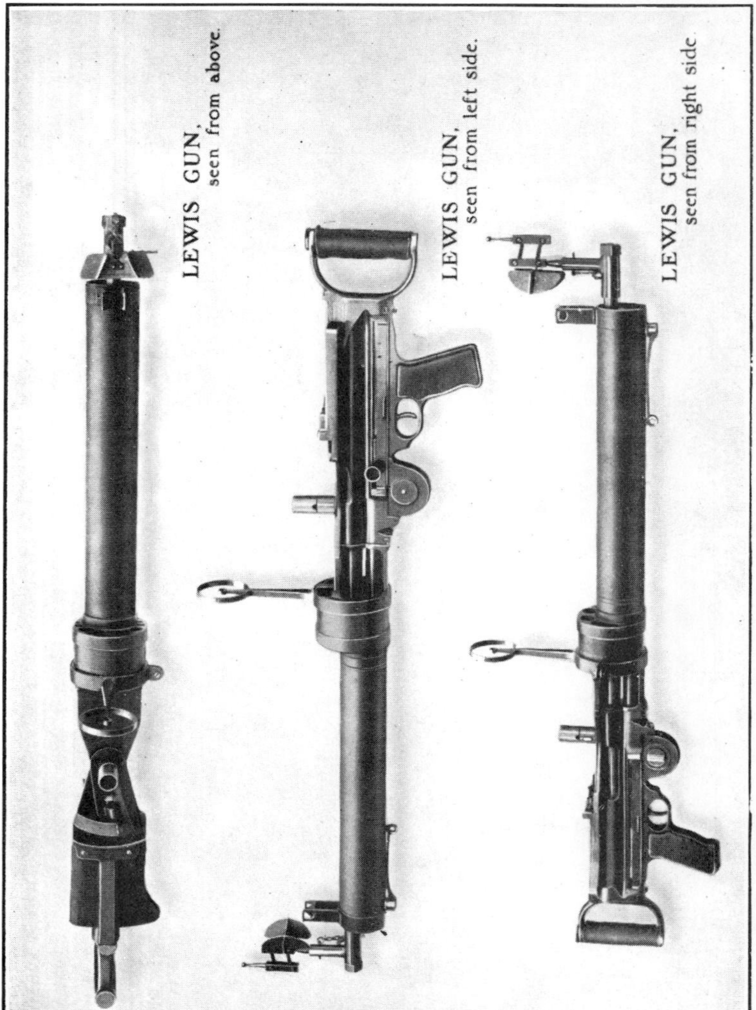

LEWIS GUN, seen from above.

LEWIS GUN, seen from left side.

LEWIS GUN, seen from right side.

HANDBOOK ON THE LEWIS GUN.

I. GENERAL DESCRIPTION.

Weight:

Ground gun (Mark I), about 26 lbs.
Aerial gun (Mark II), about 18½ lbs.
The difference in weight is mainly due to the fact that the Mark II gun has no radiator, and that the barrel casing is smaller.
Aerial gun (Marks II* and III), about 17 lbs.

Method of operation:

The gun is worked automatically by two forces :

1. The pressure of the gas generated by the explosion of the charge;
2. The return spring.

Cooling system:

The gun is cooled by air, in both the ground and aerial patterns. In the former case air is sucked through the radiator by the creation of a vacuum each time a shot is fired. In the latter case a stream of air is driven along the exterior of the barrel by reason of the rapid flight of the aeroplane.

GENERAL DESCRIPTION OF PARTS.

The gun is divided into two portions :—

(1) Stationary portions.
(2) Moving portions.

Stationary Portions.

These consist of :—

(*a*) The Barrel Group.
(*b*) The Body Group.

(a) *Barrel Group* :—

1. **Barrel.** Calibre ·303 in., number of grooves, 4; twist of rifling, right-handed.

It is threaded at the muzzle to take the *Barrel Mouthpiece,* **which has a left-handed thread to prevent it from working loose, or becoming detached from the barrel during firing.** In the Mark II gun this thread is utilized to take the nut which secures the *Norman Vane Foresight.*

A *Gas Vent* is bored in the barrel 4 in. from the muzzle to allow the gas to pass into the gas chamber. Around this vent, in the thickness of the barrel is a conical recess with a flat base.

On the top of the barrel, near the rear end, is a *Stud* **which fits in a slot in the radiator, ensuring that the gas vent is in position to receive the gas chamber.** In the Mark II gun, this stud lies between two positioning studs on the under side of the flat surface at the rear end of the barrel casing.

Behind this stud is a *Collar* which determines the position of the barrel in the casing.

Behind the collar a *Square Thread* is cut for attachment to the body.

On the rear face of the barrel is the *Cartridge Rim Stop,* a projection which coincides with the cut-away portion of the bolt face and extractors when the breech is closed. By this means the rim of the cartridge is held away from the barrel, so that the extractors are able to grip it, while at the same time the base of the cartridge is effectively supported. The upper part of the entrance to the chamber is slightly sloped, to direct the nose of the bullet downwards.

Surrounding the barrel is the *Radiator,* **which is of aluminium, has projecting flanges, and is split longitudinally. It assists in dissipating the heat of the barrel. At the forward end, on the under side, is a recess for the gas chamber. At the rear end, on the upper side, is a flat which corresponds with another flat on the radiator casing to ensure alignment.**

2. **Radiator Casing. Around the barrel and radiator is the** *Radiator Casing,* **which consists of front and rear portions, connected by a** *Clamp Ring.* **The rear portion has a hole cut in it to allow for the entrance of the gas regulator, and a small hole to take the stud of the gas regulator key. The rear face is drilled to admit and hold in position the barrel and gas cylinder. There is also a small hole drilled for the body locking pin. The front portion is tapered, and extends beyond the barrel mouthpiece. On both portions are collars which take the clamp ring, slots being cut to accommodate the stud on it. The clamp ring is turned up at both ends to form a foresight protector, and the inner portion on the right side forms the foresight block, and is dovetailed to receive the foresight. The clamp ring is held in position by the** *Clamp Ring Screw.* **On the upper side of the radiator casing and at the rear, is a flat surface, the casing being cut away immediately in front for cooling purposes.**

In the Mark II aerial gun the casing is in one piece, known as the *Barrel Casing.* There is no radiator, and the casing is of smaller diameter, except at the rear end, where it is of the same size as the ground gun. Near the front end are two holes, one to take the gas regulator and one to take the stud on the gas regulator key. A small hole is drilled in the rear face to take the body locking pin. At the rear end are two bands, one of which takes the *Mounting Yoke,* and the other the *Ring Sight.*

3. **Barrel Sleeve.** In the aerial gun the barrel is supported at the front end by a *Barrel Sleeve.* This fits the casing exactly, but is fretted for cooling purposes. Its body is cut away to allow the barrel band to be inserted.

4. **Barrel Band.** Fitting closely round the barrel, and over the gas vent, is the *Barrel Band.* Being tapered, its correct placing is ensured by keeping to the front the small letter " F " stamped upon it. The band is bored and threaded to receive the gas chamber.

5. **Gas Chamber.** This may be one of two patterns :—

(*a*) Mark I type screws into the barrel band by means of a *Nipple* chamfered away at the end to fit tightly into the conical recess in the barrel. Its rear face is threaded to receive the gas cylinder. The chamber is hollowed out and threaded to take the gas regulator. On its circumference there are two wings which provide a grip for the spanner.

(*b*) Mark II type embodies :—

(i) A ring which fits round the barrel and takes the place of the barrel band in the Mark I pattern.

(ii) A vertical seating threaded internally to take a screwed plug.

(iii) A horizontal seating into which the gas cylinder screws.

The *Screwed Plug* has a left-handed thread, and is recessed at the top to take the barrel mouthpiece spanner. It has eight holes drilled through the cylindrical portion, and is threaded internally to take the gas regulator. The plug is kept in position by a *Keeper Screw.* The holes and recesses are so arranged that when the keeper screw is in one of the recesses, one of the holes is opposite to the gas cylinder.

6. **Gas Regulator.** Screwed into the gas chamber is the *Gas Regulator,* the object of which is to regulate the amount of gas admitted to the gas cylinder. For this purpose it has two holes of different size, either of which can be placed opposite the hole in the rear face of the gas chamber. The position of these holes is indicated by the letters L and S (meaning large or small) stamped on the exposed end of the regulator.

7. It is held in position by means of the **Gas Regulator Key,** for which a square hole is cut through it. The key is held in place by a stud at its rear end, which enters a hole in the casing, a loop being provided immediately above to enable the point of a bullet to be used for springing the key out of the hole.

8. **Gas Cylinder.** This is a tube shaped at the rear end to admit the rack on the piston rod. Its forward end screws on to the gas chamber.

Mark I and Mark II cylinders differ in their manner of attachment to the gas chamber, and can only be used with Mark I and Mark II gas chambers respectively.

NOTE.—Great care should be taken to preserve the threads and to prevent the flattened portion from being damaged by the rack in stripping.

(b) *Body Group* :—

1. **Body.** This has a square thread for attachment to the barrel.

Bottom. Underneath the body is the *Housing for the Body Locking Pin,* in which lies the dumb-bell shaped *Body Locking Pin* that ensures and maintains the correct relative placing of body and barrel. At the side is a groove for the insertion of the nose of a bullet to withdraw the pin when stripping.

Immediately to the rear is the *Pinion Hinge Pin,* on which the pinion casing is hooked. Openings are cut to allow the pinion, plunger and sear to enter the interior of the body. There are also grooves in which the trigger group slides. Behind the grooves is a slot into which fit the lower projections of the butt cap.

Sides. On either side of the body is a slot in which the shank of the cocking-handle can travel. Each slot can be covered by a plate sprung into position and moved vertically by a thumb-piece. This plate serves as a *Safety Catch or Dust Protector,* the recesses in it holding the cocking-handle in the backward position. On the right side of the body is the *Ejection Opening.*

Top. Near to the front end is the *Magazine Post,* which is hollowed and countersunk to take the magazine catch. On the exterior of the post is a key which fits into a groove in the centre block of the magazine and near the base at the front is a saw-cut. The magazine post can be lengthened to take Mark II magazines by the addition of an auxiliary *Magazine Post.* The top of the body is slotted throughout the greater part of its length, the front of the slot taking the shape of a cartridge, while the rear portion acts as a guide to the boss on the feed arm actuating stud. Near the front end of the cartridge-shaped slot, on the left, is a small groove, which coincides with a depression in the front portion of the feed arm. Near the rear end of the cartridge-shaped slot are two projections which keep the cartridge in position until the bullet enters the chamber. At the rear end, on the right side, is a groove which guides the rim of the cartridge. On the left side of the slot lies the *Ejector Seating,* which is closed by an *Ejector Spring Cover* sprung into position, and held at its forward end by grooves cut in the body. At the bottom of the seating is a hole into which fits the stud on the under side of the ejector.

The *Ejector* is hook-shaped and is pivoted on a stud on its under side. It is slightly cranked, in order that, while keeping its head in line with the slot in the face of the bolt, its tail may be in position to be actuated by the lug on the bolt.

At the rear of the body is a shallow channel in which moves the stud on the tail of the feed arm.

The rear part of the top of the body has running round it a raised portion which is flanged and under cut in places, to retain the body cover in position. The curved portion of this, at the front on its left side, acts as a stop to the forward movement of the body cover. At the front end, on its right side, is a small flat portion, which acts as a guide for the cartridges. The extreme rear part of the top of the body forms a bridge, above which is a raised platform, which gives strength and helps to position the body cover.

Interior. Inside the body are drilled the *Bolt-way* and the *Piston-way.* The bolt-way has four *Cruciform Guide-ways,* or grooves in the form of a cross, which admit the lugs on the bolt and on the feed arm actuating stud. At the forward end of the bolt-way are the *Locking Recesses,* in which the resistance lugs on the bolt lie when the breech is closed. Near the rear end of the bolt-way are recesses in which the interrupted flanges of the butt cap engage. On the left of the bolt-way, slots are cut to allow the head and tail of the ejector to project alternately into it. The piston way is flat-bottomed to take the rack.

2. **Body Cover.** This is shaped to fit on the top of the rear portion of the body. At its front end is a projecting *Tongue*, on the under side of which is the seating for the cartridge guide spring.

The Cartridge Guide Spring is held in position by an under-cut recess, and has a stud which fits in a hole in the tongue. The left leaf is turned over. The right leaf is in the form of a flat spring turned in and positioned at its free end by the left leaf: it is hinged, and operated by another flat spring.

Behind the tongue is the *Housing for the Stop Pawls*, of which there are two, pivoted on studs. The *Left Stop Pawl* (marked 1) is the nearer to the top of the housing. The *Right Stop Pawl* (marked 2) has an enlarged head which brings it on a level with the under side of the body cover, and also prevents undue friction between the pawls themselves. Each stop pawl has a short tail which bears against the sides of the body cover and limits its movement. When double magazines are used, the pawls need to be bevelled to avoid the lowest rivets on the magazine. Behind the pawls is the *Stop Pawls Spring*, a flat spring which tends to push their free ends forward. On its rear face is a stud by which it is positioned. At its left end a small portion is turned over to grip the right stop pawl, so holding both stop pawls down on their studs. At the back of the housing, is a hole which accommodates the stud on the stop pawls spring. Behind this hole is a groove for the insertion of the nose of a cartridge. For lightness and the reduction of friction, the body cover is milled out, with the exception of two *Ribs*, which strengthen the cover and also act as a bearing surface for the feed arm. The body cover has round it a raised portion which corresponds with that on the body. At the rear end this raised portion fits over the raised platform on the body.

3. **Pinion Casing.** This is hollowed out to contain the pinion and return spring. At the front end is a hook which connects it to the body; this is cut away to accommodate the body locking pin, and slotted to allow the casing to be removed when the pistol grip is drawn back. At the rear end is the *Pinion Pawl*, which has two arms set at right angles, and is pivoted. One arm is actuated by a U spring bearing against it from behind; the other arm projects from the casing. When the trigger group is not in position, the former engages in the pinion to prevent the spring from unwinding. When the trigger group is in place, the projecting arm of the pawl is lifted so that the other arm is withdrawn from the pinion, leaving the pinion free to revolve. The casing is drilled to take the *Tension Screw*, and has shoulders which prevent that screw and the hub from turning when the pinion and return spring revolve. Internally it has on one side a shallow recess which accommodates a circular projection on the pinion.

The *Pinion* is a hollow casing which forms one half of a drum, and is bored centrally to take the tension screw. On its outer circumference are 52 teeth, while its inner circumference has a recess into which a corresponding projection on the spring casing enters, causing the two parts to move as one.

The *Spring Casing* contains the *Return Spring*, which is flat and coiled and has its outer end anchored by two studs which fit into two slots in the rim of the casing. The inner end of the spring is turned over to fit into a slot in the *Hub*. The latter is bored and threaded to take the tension screw.

4. **Trigger Group.** This is held in the grooves on the under side of the body. It is a frame which is channelled at the top to contain the plunger, the trigger, the sear, and the butt catch. The lower part of the frame forms the *Trigger-Guard* and *Pistol-Grip*. Externally it has grooves cut along the top edge, while at its front end is a recess for the projecting arm of the pinion pawl. The *Plunger* is a hollow cylinder with a slot to take the front end of the trigger. It contains the spring by which the trigger is actuated, and serves to position the pistol grip frame when the spade grip is removed. The *Trigger*, which is pivoted on an axis pin, has two arms, the forward one of which engages with the plunger, while the other has at its rear end a claw.

The *Sear* is also pivoted on an axis pin and has two arms, the front one engaging with the claw on the trigger, while the rear end of the other forms the sear *Nose*.

At the rear end of the pistol grip frame is the *Butt Catch*, which is operated by a spiral spring and is secured by a fixing-pin.

5. **Spade Grip.** This is fitted to the rear end of the body and comprises the *Butt Cap* and a *Handle*. On the cap is a stud with four cruciform lugs by which the spade grip is attached to the body. Below this stud is a projection with a recess into which the butt catch fits. The handle is covered with leather and hollowed out to form a reservoir for oil, this reservoir being closed by a milled-head screw to which is attached a camel hair brush.

Moving Portions.

1. **Piston Rod.** This is in two parts, joined by a loosely fitting thread secured by a pin, an arrangement designed to compensate for any want of alignment between the cylinder and the body. The piston *Head* is cupped, and has four *Rings* with the same number of *Grooves* ; these tend to hinder the escape of gas, to clear away fouling, and to reduce friction. Under the rear portion of the piston rod is a *Rack* having 60 teeth. Behind the rack is a *Bent* which engages with the nose of the sear. On the top of the piston rod is the *Striker Post*, which is drilled for the *Striker*, the latter being secured by a fixing pin; the striker post is cammed at the left front and right rear side, so as to work in conjunction with the similar cam surfaces on the slot in the bolt.

The rear end of the piston rod is cut away at the sides for lightness and the reduction of friction. Behind this cut-away portion is a slot which takes the shank of the cocking-handle.

The *Cocking-handle* has a flat *Shank* which fits into the slot in the piston rod. On this shank are two shoulders which secure the cocking-handle in the slots in the side of the body. The *Head* of the cocking-handle is cupped for the sake of lightness.

2. **Bolt.** This is a hollow cylinder. Its face is pierced by the *Striker Hole* and has a rim which supports the base of the cartridge, gaps being cut to accommodate the extractors, the cartridge rim stop, and the head of the ejector.

The two *Extractors* are flat springs, each of which fits into a recess or *Extractor Seating* cut in the outer surface of the bolt. At the front there is a claw which grips the rim of the cartridge, and a stud which sinks into a hole in the seating. The rear end is shaped to fit into a groove which positions it securely. (*Note.*—The extractor is now made in two pieces.) Lengthwise in the bolt is cut a *Cam Slot* in which the striker post travels. At the rear end of the bolt are four *Cruciform Resistance Lugs* which fit the *Cruciform Guide-ways* in the body. These are all rounded off at the right rear corner to facilitate their rotation into the locking-recesses. The top lug is flattened to allow free movement of the feed arm, while the left one is slightly cut away to admit of free movement of the tail of the ejector. The bolt is threaded internally at the rear end, and its rear face has two shoulders which act as stops for the feed arm actuating stud.

3. **Feed Arm Actuating Stud.** This is threaded externally to screw into the rear of the bolt, and has two square stops to correspond with the two on the bolt. It has four *Cruciform Guide Lugs* which also fit the cruciform guide-ways. On the topmost lug is the *Boss of the Feed Arm Actuating Stud*, a heart-shaped stud which works in the rear part of the slot in the top of the body and operates the feed arm.

4. **Feed Arm.** This has at its front end a hole which passes over the magazine post on which it pivots. A hinged *Latch* secures the feed arm to the magazine post by engaging in the saw-cut on its base. A slot called the *Feedway* is cut in the feed arm to allow the cartridges to pass from the magazine to the body. A small depression and turned-up portion of the left acts as a *Bullet Stop ;* This keeps the cartridge in position during the movement of the feed arm to the left. Slightly further back is the *Cartridge Stop*, which prevents the cartridge from jumping up after leaving the cartridge guide spring. The raised portion forming the cartridge and bullet stops is interrupted by a small cut-away part which allows clearance for the separating pegs of the magazine. Further to the rear are three studs. The largest is the *Feed Pawl Spring Stud*, which

positions one end of the *Feed Pawl Spring*. The next in size is the *Feed Pawl Stud,* which serves as a pivot for the feed pawl. The smallest is the *Feed Pawl Stop Stud,* the function of which is to stop the forward movement of the feed pawl. The *Feed Pawl* has a slot in which its spring lies, and its under side is recessed to accommodate the other end of the spring, which is positioned by the *Feed Pawl Spring Retaining Stud.* The *Tail* of the feed arm is curved and grooved on its under side to receive the boss on the actuating stud. At the end of the tail is a positioning stud which engages with the top lug of the bolt when the latter reaches its limit of movement to the rear, thus holding the feed arm in position until the boss on the actuating stud again enters the groove in the tail of the feed arm.

CONVERSION SET FOR SPEEDING UP LEWIS GUN.

In order to make Lewis Guns more certain of functioning at high altitudes with the double magazine, and also to increase the rate of fire, Conversion Sets are fitted, consisting of the following components :—

(1) Mark 1* Barrel.
(2) Mark 1* Gas Regulator.
(3) Plug and Volute Spring.

Mark 1* *Barrels* are special to the speeded-up Guns and have near the muzzle a flat on which they are marked 1* A.S. They have a larger gas hole but are otherwise the same as a Mark 1 Barrel.

Mark 1* *Gas Regulators* are also special to the Speeded-up Gun and are reduced in internal capacity by the insertion of a plug, shaped to guide the gases straight into the gas cylinder. A slot is cut in the underneath surface for easy identification, and the Regulator is marked 1* A.S.

The *Plug* is a small steel disc, with one flattened and one rounded face. It is fitted into the rear end of the piston rod, the flat face to the rear.

The *Volute Spring* is a strong steel spring made in about 4¼ flattened coils. It is inserted in the hole in the back of the piston rod, behind the plug, so that its smaller end protrudes to the rear and forms a buffer between the piston rod and the spade grip.

* * * *

GENERAL DESCRIPTION (MARK II* GUN).

This type of gun differs from the Mark II pattern in the following details :—

There is no barrel sleeve.

There is no barrel casing.

The gas cylinder is protected by a *Shield* which, at its front end, has two projections bored to receive a clamping-bolt secured by a nut. On the under side of the shield, at this end, is a projection hollowed out to receive the stud on the gas regulator key. At the rear end of the shield are two right-angled lugs bored to admit the set-screws by which it is fastened to the distance piece. Below these lugs are two cut-away portions to accommodate the heads of screws connected with the distance piece.

The gun is fitted with the " B-W " Swivelling Distance Piece, which comprises three principal parts : Front Plate, Back Plate and Swivelling Band.

The *Front Plate* is a flat circular fitting about 2¾ ins. in diameter, and ¼ of an inch thick. In this spaces are provided to admit the barrel and gas cylinder. On the face is a bossed portion having a key-way which houses the key on the barrel when the distance piece is assembled on the gun. The rear face of this plate is flat, with a rebate cut round its circumference ; upon this, when the distance piece is assembled, the flanged portion of the swivelling band is allowed to rotate.

The *Rear Plate* is similar to the front plate, and is provided with openings for the barrel and gas cylinder. In this case, however, the rebate is to the front. Protruding from the rear surface of this plate is a small portion of screw thread which allows the distance piece to be assembled to the body group and locked by the body locking pin.

The Swivelling Band consists of a flanged ring $2\frac{3}{4}$ ins. in diameter. At the base, protruding to the front, are two lugs designed for use with the standard trunnion forks.

GENERAL DESCRIPTION (MARK III GUN).

This type of gun possesses the following characteristics :—

The barrel is of the Mark I* type described above.

The gas regulator is also of the Mark I* type.

There is no barrel sleeve.

There is no barrel casing.

The gun has a gas cylinder shield identical with that described above, except that at the rear end the screws of the distance piece are accommodated by small longitudinal recesses which take the place of the cut-away portions described in the Mark II* Gun.

The heads of the feed and stop pawls are bevelled off at the top.

The safety catches are marked respectively " Left I* " and " Right I*," and have cocking-handle slots at the front as well as at the rear.

The shoulders on the pinion casing, which position the tension screw, are set wider apart than in the earlier types.

On the butt catch is a thumb-piece for releasing the catch from the recess on the projection at the base of the butt cap.

GENERAL DESCRIPTION: THE MAGAZINE.

The magazine comprises the following parts :—

1. A circular **Pan** with 25 rectangular *Indentations* and 25 *Corrugations* on the rim. Inside the indentations plates are riveted to hold the bases of the cartridges. The pan is strengthened by 25 angular plates riveted to the rim and to the top of the pan. In the centre of the top is cut a circular hole, round the edge of which are 25 recesses. Round this hole, on the under side, is riveted a ring carrying 25 *Separating Pegs*, which help to hold the cartridges in position in the magazine.

2. A **Disc.** This serves as a cover to the hole in the top of the pan. Round its edge is a flange which ensures its correct fitting. Round the disc are bored six holes to take the bolts by which it is secured to the centre block. The disc is divided into two equal parts by a raised portion (cut away in the middle) which accommodates the magazine catch. In order to facilitate the correct placing of the magazine on the magazine post, one half of the disc is covered with white paint.

3. **Centre Block.** This is made of aluminium, and is cylindrical in shape. Through the middle, from top to bottom, is a hole to fit the magazine post, this hole having a key-way which accommodates the key on the magazine post. Round the centre block are bored six holes for the insertion of the bolts by which it is bolted to the disc. Round the sides are cut four grooves, each of which is connected to the one below it by a sloping portion, the effect being to produce one continuous channel. In this the noses of the bullets are accommodated, and by means of the sloping portions the cartridges are gradually brought down to the gun as the pan revolves. At the lowest extremity of these grooves is the *Lip* of the centre block, which forms the point where the cartridges commence to leave the magazine. Opposite the lip is the *Slope* of the centre block by means of which the cartridge is forced out of the magazine when the pan rotates.

4. **Magazine Catch.** This is a strip of metal having two arms, together with a projecting under-portion shaped to form a hook. The longer arm has at the end a *Nib*, which is intended for engagement in the recesses in the top of the pan. On the top of the catch is mounted a small block of metal bored transversely to accommodate the screw which attaches it to the magazine release. The catch fits under the raised diameter of the disc, the small block then protruding above the disc.

5. **Magazine Catch Spring.** This is a small spiral spring which actuates the magazine catch. It is seated between one end of the block on the magazine catch and the projection specially made for it on one of the magazine handle brackets.

6. **Magazine Release.** This is a metal strip turned over into a rectangular shape. At its inner end the bottom is cut away to accommodate the block on the magazine catch, the release being here bored transversely to take the screw which attaches it to the catch. At its outer end it is turned up, flattened, and milled to form a grip by which it is actuated.

7. **Magazine Handle.** This consists of three portions: a leather strap, by which the magazine is lifted, and two metal brackets to which the strap is attached, and which are themselves bolted to the disc. These brackets have a raised middle portion which fits over the magazine release and the raised portion of the disc. That on the right of the magazine has a projection below which serves as a seating for one end of the magazine catch spring.

[The magazine holds 97 cartridges. When empty its weight is $3\frac{3}{4}$ lbs.; when full, 8 lbs. 14 ozs.

METHOD OF INSTRUCTION.

The teaching of this branch of the work calls for a considerable effort of memory. It should, therefore, be taken in short sections, each of which should be thoroughly understood before passing on.

The general description should be brief, the main idea being to call the attention of the pupils to the peculiarities of the various parts, with a view to their subsequent understanding of the mechanism of the gun.

It has been necessary to introduce in a few places parts not yet taught, for the clearer explanation of the particular portion under consideration. On the rare occasions when this is necessary, the instructor should refer to the unknown part without giving its name. The stage at which each new name should be introduced is indicated by *italics*.

A short general account of the gun should first be given, so that the pupils may get an elementary idea of what the gun is and what it can do, and that their interest may be aroused. This description should not take more than a few minutes, and should touch on the following points:—

Name: ·303 in. Lewis Machine Gun, and why so called.

Weight: (As on p. 7).

Method of operation: (As on p. 7).

Explain briefly:

(*a*) How the gun is loaded and fed.
(*b*) How gun is fired.
(*c*) How gas operates piston and other moving portions.
(*d*) How return spring operates piston and bolt.
(*e*) How gun is cooled.

NOTE.—This should be general, without the mention of names. Avoid talking of things that cannot be seen. Detailed explanations will come later when the gun is stripped.

Advantages:

(*a*) Lightness.
(*b*) Compactness.
(*c*) Reliability.
(*d*) Rapid rate of fire.

Disadvantages:

Smallness of magazine supply.

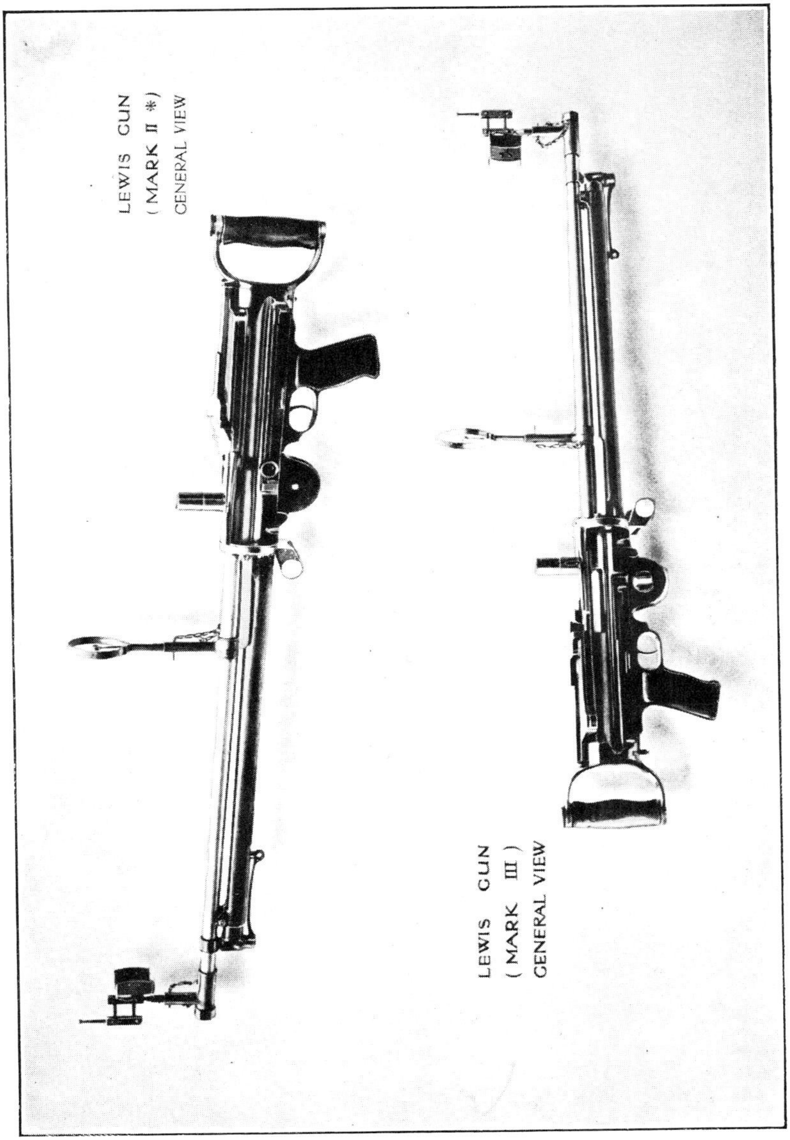

LEWIS GUN
(MARK II *)
GENERAL VIEW

LEWIS GUN
(MARK III)
GENERAL VIEW

3

Vickers .303 in Gas Operated Mk. I Gun

CONTENTS

	Page
Leading particulars	7
List of illustrations	8
Introductory	9

CHAPTER 1

Description of the gun

Vickers ·303 in. gas operated MK. I, No. 1 gun	*Para.*
General	6
The stationary portions	7–38
Barrel	8
Gas block	9
Flash eliminator	10
Foresight bracket	11
Rear sight bracket	12
Front magazine catch	13
Body	14
Gas cylinder	21
Barrel strap	22
Rear magazine catch	23
Locking shoulder	24
Ejector	25
Cocking handle	26
Cocking handle slide	27
Cocking handle catch	28
Cocking handle lug	29
Body extension	30
Buffer	31
Trigger mechanism	32
Trigger, Mk. I	33
Trigger rod with trigger spring sleeve and cap	34
Sear and sear buffer	35
Sear catch	36
Spade grip	37
Safety catch, Mk. I	38
The moving portions	39–47
Piston	40
Return springs and return springs rod	42
Breech block	43
Firing pin	45
Feed-piece	46
Extractor	47
Vickers ·303 in. gas operated Mk. I, No. 3 gun	47A
Vickers ·303 in. gas operated Mk. I, No. 2 gun	
General	48
Shoulder-piece	50
Trigger guard frame	51
Trigger, Mk. II	52
Trigger bar	53
Safety catch, Mk. II	54
Vickers ·303 in. gas operated Mk. I, No. 4 gun	54A

CHAPTER 2

Mechanism or action of the gun

	Para.
General	55
Cocking by hand	56
Forward movement	57
Backward movement	59
Action of the firing mechanism	62
Safety arrangements	65

CHAPTER 3

Miscellaneous

Deflector bag	66–68
Frame	67
Bag	68
Magazine No. 1, Mk. I	69–75
Magazine body	70
Top plate and main spring	73
Bullet support with platform and spring	74
Loading lever	75
Action of magazine	76
Magazine No. 2, Mk. I	77–84
Magazine body	78
Follower	81
Top plate and main spring	82
Loading lever	84
Care and cleaning	85

CHAPTER 4

Loading, unloading, stoppages and immediate action

Loading and unloading by hand	91–137
To load the No. 1, Mk. I magazine	91
To tension the mainspring	100
To remove tension from the main spring	102
To unload the No. 1, Mk. I magazine	105
To load the No. 2 Mk. I magazine	109
To tension the mainspring	120
To remove tension from the mainspring	123
To unload the No. 2, Mk. I magazine	126
To load the gun	129
To unload the gun	134
Stoppages and immediate action	138–142
Stoppages	138
Immediate action	140
Immediate action table	142

LEADING PARTICULARS

Length of gun overall with flash eliminator	3 ft. 4 in.
Length of barrel	1 ft. 8 in.
Weight of gun (approx.)	
(i) Mk. I, No. 1 gun	$19\frac{3}{4}$ lb.
(ii) Mk. I, No. 2 gun	20 lb.
Rifling	
(i) Type	Left hand
(ii) Twist	1 in 10 in.
(iii) Number of grooves ..	5
Ammunition used	All types of Mk. VII S.A.A.
Method of feed	Magazine
Capacity of magazine	(i) 60 rounds (ii) 100 rounds
Speed of fire (approx.)	
(i) Mk. I, No. 1 gun	950 rounds per min.
(ii) Mk. I, No. 2 gun	950 rounds per min.
(iii) Mk. I, No. 3 gun	950 rounds per min.
(iv) Mk. I, No. 4 gun	700 rounds per min.
Marks on the gun	
(i) On the side of the breech casing	Nature, mark and number of the gun
(ii) On the barrel	Proof mark Calibre Year of manufacture (last two figures) Number of the barrel
(iii) On the breech block ..	Number of the breech block
(iv) On various components ..	Viewer's marks Broad arrow Maker's trade mark
Sights	Separate store
Mounting	
(i) Air service	Turret or free installations
(ii) Ground service	Tripod

INTRODUCTORY

1. The Vickers gas operated gun has a high rate of fire combined with a smooth action and is fully automatic. The mechanism is simple, the working parts are few in number and the gun can be stripped for cleaning in less than 1 minute, while the barrel can be readily removed and exchanged without stripping any other part of the gun. One special feature of the gun is that it has no external moving parts. Ammunition is fed from a spring-loaded drum shaped magazine, situated on the top of the gun, which can be removed and exchanged in a few seconds.

2. After the first round has been loaded and fired, the necessary motive power to operate the gun is obtained by deflecting a portion of the propelling gas, through a hole in the barrel near the muzzle, on to a piston. The piston is driven to the rear, compressing the return springs and unlocking and drawing back the breech block. This, as it moves to the rear, ejects the empty cartridge case from the chamber into a deflector bag on the side of the gun. The piston, and consequently the breech block, are held in the rear or cocked position by a sear until the firing control mechanism is operated. The operation of the firing control mechanism releases the sear, allowing the return springs to react and drive forward the piston and the breech block which pushes a round from the magazine into the chamber. After the breech block is locked, the piston continues to move forward and strikes a floating firing pin in the breech block to fire the round. So long as the firing control mechanism is operated, the gun should be automatically loaded and fired until the ammunition in the magazine is expended.

3. The firing control mechanism may take one of several forms. In the original design of gun as described in this handbook, firing is by means of a trigger situated at the rear, but when mounted in aircraft a hydraulic, pneumatic or an electrical device may be employed. The device takes the form of a sear release unit, replacing the spade grip and Mk. I trigger and is described in A.P.1641E, Vol. I. With some installations a trigger operated by a bowden wire may be employed in place of the Mk. I trigger. The Mk. I, No. 2 gun is fitted with a Mk. II trigger which is connected to the Mk. I trigger by a trigger bar.

4. The Mk. I gun can be readily adapted for air service or ground service use by fitting the appropriate No. 1 gun or No. 2 gun attachments.

5. Air flow over the barrel and flash eliminator provides the necessary cooling for these parts of the gun. When mounted in aircraft, arrangements for heating may be made to prevent freezing of the moving parts of the gun.

CHAPTER I

DESCRIPTION OF THE GUN

VICKERS ·303 IN. GAS OPERATED MK. I, NO. 1 GUN

General

6. The gun may be divided naturally into two parts for the purpose of description.

(i) The stationary portions.

(ii) The moving portions.

THE STATIONARY PORTIONS

7. The stationary portions consist of the following groups.

(i) Barrel with gas block, flash eliminator, sight brackets and front magazine catch.

(ii) Body with rear magazine catch, ejector, gas cylinder and barrel retaining strap.

(iii) Cocking handle.

(iv) Body extension with buffer, trigger mechanism and spade grip with safety catch Mk. I.

Barrel (fig. 1 and 3)

8. The barrel is bored longitudinally and rifled to suit S.A. ammunition. At the rear the bore is increased in diameter to form a chamber which is coned to conform to the contour of the Mk. VII cartridge case. Between the chamber and the commencement of the rifling is a second cone which forms a lead for the bullet into the rifling. The breech end of the barrel is supported in the front of the body, being positioned by a collar and having radial projections on the top for engagement with corresponding grooves in the barrel strap, which holds it in the body. To ensure correct assembly the projections are of unequal width. The breech face of the barrel is bevelled to clear the extractor. A small gas hole or vent is drilled radially through the barrel approximately 6 in. from the muzzle, which is reduced in external diameter and threaded to take the flash eliminator. Behind the flash eliminator is fitted and secured the foresight bracket.

Gas block (fig. 3)

9. The gas block is permanently attached to the barrel and a small hole in it coincides with the vent. It is secured by a pin which engages a groove in the underside of the barrel. Beneath the hole in the gas block is a conical cavity below which the block is threaded to receive the gas plug. This plug is tapered to fit the cavity and is slotted for the gas plug port which has a curved groove on one side through which the gas passes to a hollow spigot on the rear of the gas block and then into the gas cylinder. The bottom of the gas plug is hexagonal and when assembled in the gas block the plug is secured by a split pin.

Flash eliminator (fig. 3)

10. The flash eliminator, which diminishes flash from the gun during firing, is conical in shape and is screwed to the muzzle end of the barrel. It is provided with flats on its exterior for the combination spanner, used in fitting and removing, and has a hole in a projecting lug for a split pin which prevents it from unscrewing.

Foresight bracket (fig. 3)

11. The foresight bracket fits round the barrel behind the flash eliminator and is located radially by a key which is permanently attached to the barrel. Formed on the bracket is a hollow vertical projection for the attachment of the foresight which is secured by a pin.

Rear sight bracket (fig. 3)

12. The rear sight bracket is riveted to the top of the front magazine catch bracket which is permanently attached to the barrel near the breech end. The bracket is provided with a hollow vertical projection for the attachment of the rear sight which is secured by a split pin. When a reflector sight is used this is fitted into a slot at the rear of the body.

Front magazine catch (fig. 3)

13. The front magazine catch is a spring-loaded plunger which is retained in the bracket on the top of the gun by a split pin. The catch is chamfered on the top and relieved on the underside to facilitate positioning the front catch plate of the magazine.

Body (fig. 1 and 4)

14. The body is of a general rectangular shape and is prepared at the front to form a support for the breech end of the barrel. A slot in the top and vertical grooves in the sides accommodate the barrel strap which is secured by the bolt of the mounting-yoke which passes transversely through a hole in the body. A second hole is drilled through the front for the bolt of the trigger guard frame.

15. Behind the barrel support and extending to the rear of the body a square hole is formed to accommodate and guide the breech block. The right hand side is recessed to clear the extractor and is provided with an orifice to allow the empty cartridge cases to be ejected, A bracket is screwed to the body above the orifice and has a projector which engages the catch on the frame of the deflector bag.

16. A second orifice is cut in the top of the body to enable ammunition to be loaded from the magazine. To the rear of this orifice there is a projection which is prepared to accommodate the rear magazine catch and under the projection the locking shoulder is fitted and secured by a pin. On the left side of the body there is a slotted box-shaped projection which accommodates the ejector.

17. Below and parallel with the square hole there is a round hole into the front of which fits the rear end of the gas cylinder, the remainder of the hole acting as a piston-way for the rear of the piston rod. Fitted transversely through the body and towards the front are the top and bottom piston stops which project into the piston-way and form a stop for the piston at the end of its forward movement. These stops are permanently attached to the body.

18. On each side of the piston-way there are keyways which engage corresponding keys on the piston. A slot in the left hand keyway enables the cocking handle lug to engage the piston. On each side of this slot there are grooved recesses in which the cocking handle slide travels, the front being undercut for the cocking handle catch. At the rear end the grooves are relieved to facilitate assembling and dismantling the cocking handle slide. A clearance in the top of the piston-way allows a projection at the rear end of the piston to engage the breech block.

19. The rear of the body is suitably shaped to receive the body extension which is retained in position by upper and lower securing pins. These pins are held transversely by spring-loaded retaining plungers, which are housed in projections formed on the sides of the body. The plungers engage grooves, cut for the greater part of the length of the securing pins, and allow the pins to be withdrawn sufficiently to enable the body extension to be removed.

20. The bottom plate of the body is riveted in position and is prepared with two hinge lugs to enable the deflector bag to be attached to the gun.

Gas cylinder (fig. 1 and 9)

21. The gas cylinder is tubular in shape and forms a guide for the piston head as this travels backwards and forwards. The front end is enlarged to fit over the spigot on the gas block, the rear end fitting into the front of the body. The gas cylinder is a loose fit and has a number of radial holes drilled in it to allow the gas to be readily expelled as the piston moves forward.

Barrel strap (fig. 1 and 4)

22. The barrel strap is U-shaped and is grooved on the inside to engage the projections on the barrel. Two holes are drilled through the lower part of the strap, towards the front, for the retaining bolt which also secures the mounting yoke.

Rear magazine catch (fig. 4 and 8)

23. The rear magazine catch has two projections at the front which are chamfered on the top and relieved on the underside to facilitate engagement with the rear catch plate of the magazine. It slides longitudinally on the projection on the top of the body and is kept up to its work by a spring which is retained between a plug in the projection and a recess in the bottom of the catch. A lever, to enable the catch to be disengaged and the magazine removed, is pivoted between two lugs on the top of the projection, its lower end being suitably shaped to engage the front of the catch.

Locking shoulder (fig. 1, 4 and 12)

24. The locking shoulder, which supports the breech block against the shock of discharge, is housed in the top of the body and is retained by a pin. A projection on the locking shoulder is recessed for engagement with the catch on the frame of the deflector bag. The locking shoulder must not be removed except for purposes of repair in a depot.

Ejector (fig. 8)

25. The ejector is a short lever with a lug at each end and having trunnions towards the rear about which it rotates. The trunnions are partly cut away for assembly in the bayonet pattern slots in the projection, on the left hand side of the body, and form the axis of the ejector. The rear lug is engaged by a cam on the breech block as this travels to the rear, while the front lug engages and knocks the empty cartridge case from the face of the breech block as the ejector rotates. A cover of thin steel plate is dovetailed into the projection on the body to retain the ejector in position, a thumb piece being riveted to the cover to facilitate removal.

Cocking handle (fig. 1 and 8)

26. The cocking handle, the object of which is to enable the piston and breech block to be withdrawn by hand to the cocked position, is situated on the left hand side of the body. After the gun has been cocked, the handle is returned to its forward position where it is retained by a spring-loaded catch. The cocking handle is hollow and is pivoted at the front of the cocking handle slide, being held in its normal position by a small torsional spring which bears against the inside of the cocking handle and a projection on the cocking handle slide.

Cocking handle slide (fig. 8)

27. The cocking handle slide which governs the movement of the cocking handle is suitably shaped to slide longitudinally in the grooved recess in the body. It has a projection on the outside over which the cocking handle fits and carries a spring loaded catch. A slot is cut in the slide for the cocking handle lug by means of which it engages the piston.

Cocking handle catch (fig. 8)

28. The cocking handle catch is a right-angle shaped lever and is mounted on the same axis pin as the cocking handle. One arm fits inside and is engaged by the cocking handle when this is operated, the other arm being provided with a claw which engages the end of the grooved recess in the body. The spring bears between the projection on the cocking handle slide and the arm of the catch.

Cocking handle lug (fig. 8)

29. The cocking handle lug fits in a slot to the rear of the cocking handle slide and projects through the slot in the body, to engage the piston when the cocking handle is pulled to the rear. It is held in position by a pin which is retained by a split pin.

Body extension (fig. 1, 4, 5 and 12)

30. The body extension is shaped to fit the rear of the body and to accommodate the buffer and trigger mechanism, suitable holes being drilled for the various axis pins. Two holes are also drilled through the extension for its securing pins and two for the screws of the spade grip. Another hole is drilled through a projection on the underside for the attachment of the gun to the mounting or for the attachment of the trigger guard frame of the No. 2 gun.

Buffer (fig. 5 and 12)

31. The buffer against which the piston rebounds is housed in a hole drilled through the rear of the body extension in line with the piston, the rear of the hole is threaded for the buffer spring screw which is retained by a transverse pin. The buffer is cylindrical in shape and is enlarged at the rear to form a stop which limits its forward movement. It is bored from the front to accommodate and form a bearing for the rear of the return springs rod. The buffer spring is a helical compression spring and is retained against the rear of the buffer by the buffer spring screw. A steel buffer disc is placed between the spring and the screw.

Trigger mechanism (fig. 5 and 12)

32. The trigger mechanism is so designed that unless the trigger is fully released the sear will not rise and engage the bent on the piston, a spring-loaded sear catch having been incorporated for this purpose.

Trigger Mk. I (fig. 5 and 12)

33. The Mk. I trigger pivots on a pin at the rear of the body extension, the buffer screw forming a stop for it in its normal position. The top of the trigger is suitably shaped for engagement with the Mk. I safety catch, when this is set to safe, while at the bottom a hole is drilled for the pin of the trigger rod.

Trigger rod with trigger spring sleeve and cap (fig. 5 and 12)

34. The trigger rod, which transfers the motion of the trigger to the sear, is rectangular in section and is pinned at the rear to the trigger. At the front the rod is forked to engage the sear, while behind the fork a stop is riveted. The trigger spring sleeve and trigger cap are assembled on the rod. The sleeve houses the trigger spring and the cap, butting against the trigger, forms a rear bearing for the spring. When assembled in the body extension the sleeve bears against a shoulder in the former, being prevented from turning by a lug. The sear catch spring is carried on the rod in front of the sleeve and the sear spring in the lug on the sleeve.

Sear and sear buffer (fig. 5 and 12)

35. The sear is approximately L-shaped and is pivoted in a projection at the front of the body extension. The axis pin holes are elongated to allow of slight longitudinal movement of the sear when it is engaged by the bent of the piston. The movement causes the sear buffer spring to be compressed and absorb the shock of engagement. The underside of the sear bent is shaped to engage a transverse pin in the body extension as the sear moves forward, this ensures that the sear will rise to its fullest extent, and a lug at the rear of the bent engaging the body extension limits the upward movement. The sear spring is flat and as mentioned previously is carried on the lug of the trigger spring sleeve. The front of the sear is cut away to bear against the sear lever which is pivoted in the body extension and bears against the sear buffer plunger. The sear buffer spring is housed in a sleeve which is screwed into the front of the body extension and is retained against the head of the plunger by a screwed plug and pin.

Sear catch (fig. 5 and 12)

36. The sear catch, which prevents the sear from rising in the event of the trigger being eased, is pivoted in the bottom of the body extension. It has two parallel arms which pass one on either side of the trigger rod behind the stop, the arms having projections at their extremities which engage the lug on the sear when the trigger is pulled. The catch is functioned by its spring which is held in initial compression against the front of the trigger spring sleeve. When the trigger is fully released the stop at the front of the trigger rod disengages the catch allowing the sear to rise.

Spade grip (fig. 1, 4 and 6)

37. The spade grip is a hollow steel frame and fits partly over the body extension to which it is secured by two screws. It is cut away to enable the trigger to be operated and also to form a grip. Two wooden side pieces are clamped on the grip by two screws. The upper screw also forms the axis of the Mk. I safety catch which is operated through a slot in the top of the spade grip at the rear.

Safety catch Mk. I (fig. 6)

38. The Mk. I safety catch is operated by a small knurled thumbpiece which also pivots on the upper side piece screw. The thumbpiece is semi-circular in shape, hollow, and has a projection near its axis, to which a plunger is pinned. The letters S and F are engraved upon the thumbpiece. The safety catch is connected to the thumbpiece by the plunger which passes through a plunger guide pivoted on trunnions in a slot in the catch. The spring presses between a shoulder on the plunger and the plunger guide and ensures that the catch is retained in the safe or fire position.

THE MOVING PORTIONS

39. The moving portions consist of the following groups :—

(i) Piston and return springs with rod.

(ii) Breech block with firing pin, feed piece and extractor.

Piston (fig. 9 and 12)

40. The piston is a hollow steel rod having a head formed at its front end. The head is cup-shaped to retain any fouling, thus preventing its spreading over the periphery of the head, and it also has annular grooves the edges of which help to keep the gas cylinder free from fouling. The rod is enlarged in diameter near the middle to form a bearing for it in the gas cylinder. The rear of the piston has keys on each side which engage corresponding keyways in the body. In the left hand keyway is formed a groove for the cocking handle lug, and when the cocking handle is pulled to the rear this lug engages the end of the groove to withdraw the piston. On the underside at the rear is formed a bent which engages the sear to hold the piston in the cocked position.

41. On the top of the piston at the rear are two projections having inclined parallel surfaces which engage corresponding surfaces in the bottom of the breech block. The front inclined surface lowers the breech block disengaging it from the locking shoulder as the piston is moved backwards, while the other inclined surface raises the breech block to its locked position in front of the locking shoulder as the piston approaches the firing position. The rear projection has a flat on the top which supports the rear of the breech block in the locked position when the piston is right home. The front projection is shaped to form a hammer and strikes the firing pin as the piston reaches the end of its forward movement, which is arrested when a shoulder on the piston meets the stops in the body.

Return springs and return springs rod (fig. 9 and 12)

42. The return springs consist of an outer and inner spring the front ends of which bear against shoulders inside the piston. The rear ends of the springs are supported on the rod which has a shoulder near its rear end to form a bearing for the springs. The rear of the rod is supported inside the buffer against which the shoulder abuts.

Breech block (fig. 10 and 11)

43. The breech block is rectangular in shape, its underside being recessed to accommodate the front projection on the piston. It is provided with inclined surfaces to engage those of the piston. On the right hand side towards the front, the breech block is recessed and undercut to accommodate the extractor spring and a hole is drilled and threaded for the firing pin screw which also retains the extractor spring. At the front the breech block is suitably cut away to house the extractor and is shaped to receive the base of the cartridge case. A cam on the left hand side of the breech block controls the action of the ejector while at the front there is a slot to enable the ejector to operate and eject the empty cartridge case.

44. The breech block is recessed at the top towards the front to accommodate the feed piece with its plunger and spring, the axis pin of the feed piece passing transversely through the breech block. At the rear there is a projection which engages the locking shoulder in the body to support the breech block in the firing position. The breech block is drilled and bored longitudinally to receive the firing pin and spring, a small gas escape hole being drilled at an angle from the underside into the firing pin hole.

Firing pin (fig. 11)

45. The firing pin is cylindrical in shape, tapering to a point at the front and having a head formed at the rear. A shoulder towards the rear forms a bearing for the spring, the front of which abuts against a shoulder in the breech block to retain the firing pin in the withdrawn position until the head is struck by the projection on the piston. The firing pin retaining screw engages against the top of the head to retain the firing pin in the breech block.

Feed-piece (fig. 10 and 11)

46. The feed-piece is short lever pivoted in the recess in the breech block and is designed to push a round from the magazine into the chamber during the forward movement of the breech block. A plunger and spring are positioned under the front of the feed-piece to ensure engagement with the base of the cartridge case.

Extractor (fig. 10 and 11)

47. The extractor is held longitudinally by keys and keyways and laterally by a flat spring in its recess in the breech block. The spring allows the extractor to move outwards sufficiently to ride over, and engage the rim of the cartridge case in the chamber as the breech block reaches the firing position. The spring is held in its undercut recess in the breech block by the firing pin screw.

VICKERS ·303 IN. GAS OPERATED MK. I, NO. 3 GUN

47A. The Mk. I, No. 3 gun differs from the Mk. I, No. 1 gun in being fitted with remote control for use in the air. A dust cover replaces the spade grip and a special trigger with bowden wire attachment is fitted instead of the Mk. I trigger. A bracket to support the bowden cable is fitted to the underside of the body extension and secured by split pins.

VICKERS ·303 IN. GAS OPERATED MK. I, NO. 2 GUN

General

48. The Mk. I, No. 2 gun differs from the Mk. I, No. 1 gun as follows :—

(i) A shoulder-piece is fitted in place of the spade grip.

(ii) It is fitted additionally with a trigger guard frame, Mk. II trigger, trigger bar and Mk. II safety catch.

(iii) The mounting yoke bolt, which secures the barrel strap, is replaced by one having trunnions for the attachment of the gun to the tripod mounting, the mounting yoke being removed.

49. In other respects the two guns are identical and para. 6 to 47 should be referred to for information relating to those parts of the No. 2 gun which are common to the No. 1 gun. A description of the shoulder-piece, trigger guard frame and their components follows.

Shoulder-piece (fig. 2, 4 and 7)

50. The shoulder-piece body is of steel sheet and is shaped to fit over the body extension and Mk. I trigger being held in position by two screws which are secured by split pins. On the underside at the front are two projections through which passes the securing pin of the trigger guard frame. This pin is held in a manner similar to the securing pins of the body extension, the plunger and spring being housed in a projection on the side of the shoulder-piece. Guide pins are riveted inside the shoulder-piece to support the rear portion of the trigger bar. A shoulder pad, consisting of a wood block, sorbo pad and leather cover, is secured to the rear of the shoulder-piece by wood screws.

Trigger guard frame (fig. 2 and 7)

51. The trigger guard frame is of steel sheet and is in general appearance trough-shaped. It is held in position, against the underside of the gun, by a bolt which passes through the forward hole in the body and by a securing pin which passes through the rear mounting lug on the body extension. At the front a wood block is secured by two screws to form a hand grip and near the middle is a pistol grip which is provided with wooden side pieces. In front of the pistol grip is a trigger guard and above it is fitted the Mk. II safety catch. Guide pins are riveted inside the frame to support the front portion of the trigger bar.

Note.—The fitting of the trigger guard frame prevents the attachment of the deflector bag.

Trigger Mk. II (fig. 2 and 7)

52. The Mk. II trigger, which facilitates the firing of the gun from the shoulder, is connected to the Mk. I trigger in the body extension by the trigger bar. It is mounted on an axis pin in the trigger guard frame, the axis pin being secured by a split pin. A projection is formed at the top of the trigger for the attachment of the trigger bar.

Trigger bar (fig. 7)

53. The trigger bar is in two parts, the front portion being accommodated in the trigger guard frame and the rear portion in the shoulder-piece. Each part consists of two rods which pass on either side of the body

extension when the components are assembled to the gun. The rods are held rigid by distance pieces and are supported on the guide pins mentioned previously. The front portion is connected to the Mk. II trigger by a pin and has slots cut in it for engagement with the Mk. II safety catch. When the trigger guard frame and shoulder-piece are fitted to the gun, projections on the front portion of the trigger bar engage corresponding projections on the rear portion, the distance piece of which bears against the lower end of the Mk. I trigger in the body extension.

Safety catch Mk. II (fig. 2 and 7)

54. The Mk. II safety catch is held in position in the trigger guard frame by a bayonet joint. It engages the slots in the trigger bar when set to " Safe " and locks the action, but when set to " Fire " a clearance on the catch allows the mechanism to be operated. A spring steel plate having two studs at each end is secured inside the trigger guard frame by a central rivet. The studs project through the frame and hold the catch in the safe or fire position and also limit the movement of the catch and prevent disengagement of the bayonet joint.

VICKERS ·303 IN. GAS OPERATED MK. I, NO. 4 GUN

54A. The Mk. I, No. 4 gun is for use in armoured cars and differs from the Mk. I, No. 3 gun in being fitted with a Mk. II gas plug, which reduces the speed of fire, and in having the front hand grip made of hard rubber.

CHAPTER 2

MECHANISM OR ACTION OF THE GUN

General

55. Owing to the very high rate of fire of this gun all the mechanical operations are practically instantaneous, the time interval between explosions being about $\frac{1}{16}$ sec.

Cocking by hand

56. When the cocking handle is pulled to the rear the catch is released and allows the cocking handle to move backwards. The cocking handle lug engaging the piston causes this to be drawn to the rear to unlock and carry with it the breech block. At the same time the return springs are compressed. If the trigger is released the piston bent will ride over the sear and depress it, but the sear spring reasserting itself will cause the sear to rise and engage the bent on the piston to hold it in the cocked position.

Forward movement. (Assume a loaded magazine to be in position) (fig. 12)

57. When the trigger is pressed the sear is disengaged from the piston bent and the return springs return the piston and breech block to the firing position. As the breech block moves forward the feed-piece engages the base of a cartridge in the magazine and pushes the round forward. The bullet is deflected downwards by the bullet guide on the magazine into the chamber and the final forward movement of the breech block rams home the round, the extractor riding over and engaging the rim of the cartridge case.

58. The piston continues its travel after the breech block is right home and the inclined surface on the rear projection, engaging the corresponding surface on the breech block, causes the rear of the breech block to rise and engage in front of the locking shoulder. The breech block is held in front of the locking shoulder by the flat at the rear of the piston, which locks it against the force of the explosion. During the final movement of the piston, the front projection strikes the firing pin to fire the round, the piston coming to rest against its stops in the breech casing.

Note.—The locking shoulder and piston stops indicated in fig. 12A are not the actual fittings but have been drawn on the figure to show their relative positions.

Backward movement. (Assume the gun to have just fired) (fig. 12)

59. After the bullet has passed the gas vent in the barrel a portion of the propelling gas escapes into the gas block and is deflected by the gas plug port on to the head of the piston driving it to the rear and compressing the return springs. The piston moves about $\frac{3}{10}$ in. before the inclined surface on the front projection engages the corresponding surface on the breech block to lower this and disengage it from the locking shoulder in the body. Whilst this movement is taking place the rear of the breech block is supported by the horizontal flat on the rear of the piston, which ensures that the breech block will remain locked to the gun until the bullet has left the muzzle, and the firing pin is withdrawn into the breech block by its spring.

60. The breech block is now carried to the rear by the piston and, as the breech block moves away from the barrel, the extractor extracts the

empty cartridge case from the chamber. The cam on the left hand side of the breech block causes the ejector to rotate sharply and engage the base of the empty cartridge case which is ejected, through the orifice in the breech casing, into the deflector bag. As the feed-piece passes under the next round in the magazine it is deflected downwards, being returned into position by its spring, ready to engage the base of the cartridge when the breech block moves forward.

61. The backward movement is arrested when the piston strikes the buffer in the body extension. The piston rebounds and if the trigger is kept pressed the cycle of operations is repeated until the last round in the magazine has been fired. Should the trigger be released before the ammunition is expended, the sear will rise and engage the bent on the piston, as this begins to move forward, to retain the piston and breech block in the cocked position. Thus the gun does not cease firing with a live round in the chamber.

Action of the firing mechanism (fig. 12)

62. When the Mk. I trigger is operated, either direct or through the medium of the Mk. II trigger, it pivots on its axis pin and the lower end pushes forward the trigger rod which engages the sear. The sear is caused to rotate until the bent is drawn below the level of the projection on the body extension and clear of the piston bent. The movement of the trigger rod compresses the trigger spring and also allows the sear catch spring to react. The latter spring bearing against the sear catch causes this to rotate and engage the lug on the rear of the sear and retain it against the pressure of the sear spring.

63. The trigger is returned to its normal position by its spring and draws back the trigger rod when pressure is released. The sear, however, does not rise immediately as it is retained by the sear catch. As the trigger rod nears its normal position, a stop on its inner end bears against the sear catch and disengages this from the sear, which rises under the influence of its spring. When the piston rebounds off the buffer the bent strikes the sear and drives it forward. The sear bearing against the sear lever causes this to swing forward against the plunger and compress the sear buffer spring, which absorbs the shock of engagement and then, reasserting itself, returns the sear to its normal position.

64. If a hydraulic, pneumatic or an electrical firing control mechanism is fitted, the Mk. I trigger is removed and its action is simulated by a plunger in the device which engages the trigger rod, the action of the gun mechanism being the same as already described.

Safety arrangements

65. The gun cannot be fired until after the breech has been locked for the following reasons :—

(i) The design of the mechanism is such that the piston must raise the rear of the breech block in front of the locking shoulder before it can move forward independently to strike the firing pin.

(ii) If the breech block is not raised the piston will jam.

(iii) The firing pin is retained within the breech block by its spring until it is struck by the piston.

CHAPTER 3

MISCELLANEOUS

DEFLECTOR BAG (fig. 1 and 13)

66. The deflector bag which retains the empty cartridge cases as they are ejected from the gun consists of two main parts :—

(i) The frame.
(ii) The bag.

Frame

67. The frame is of sheet steel and is open at one side and at the bottom. A bracket is riveted below the side opening to enable the frame to be hinged to the lugs on the body, the hinge pin being secured by a split pin. At the top there is fitted a spring-loaded catch by means of which the frame is held against the ejection orifice of the body. Two spring steel baffle plates are riveted inside the frame, one on the side opposite to the opening and the other at the rear. The object of these plates is to deflect the empty cartridge cases downwards into the bag. Flanges are formed at the bottom of the frame to engage corresponding flanges on the bag and a lug is riveted in position to engage the catch of the latter.

Bag

68. The bag is in two parts which are riveted together. The top is of sheet steel and is flanged to fit the bottom of the frame, a spring-loaded catch being provided to retain the bag on the frame. The bottom portion is of stout canvas and is provided with a U-shaped steel stiffener which is riveted to the top of the bag.

MAGAZINE NO. 1, MK. I (fig. 1 and 14)

69. The No. 1, Mk. I magazine is designed to hold 60 rounds and consists of three main parts :—

(i) Magazine body.
(ii) Top plate with separators, main spring, spring casing and bullet support with platform and spring.
(iii) Loading lever and retaining pin.

Magazine body

70. The magazine body consists of a bottom plate and band riveted together and having two catch plates situated at the front and rear on the outside. The catch plates engage the front and rear magazine catches on the gun, when the magazine is in position. Near the catch plates are two spring clips which retain the top plate when the magazine is assembled.

71. A slot is cut in the bottom plate to coincide with the orifice in the gun body. On either side of this slot there are lips which project into the orifice when the magazine is in position, and, being suitably shaped, retain the round until it is pushed out of the magazine by the feed-piece. At the front of the slot on the underside is a bullet lead which deflects the round downwards as it is pushed forward. Near the middle of the bottom plate is a flat spring which acts as a support and guide for the bullets, and in the centre is a hollow spigot or centre post to which the bush is pinned when in position.

72. Inside the band, which has a flange at the top, there is a cartridge head guide ring which forms a gallery to support the base of the cartridge cases, a cartridge guide being provided to deflect the cartridges through the slot in the bottom plate.

Top plate and main spring

73. The top plate is assembled on the top of the magazine body and is secured by the spring clips which project over its edge. The separators, which locate the rounds radially in the magazine, are riveted to the underside of the plate. A spring casing which is in the form of a hub, is also positioned on the underside of the plate and houses a clock pattern main spring. The outer end of the spring is anchored to the casing and its inner end is slotted to engage a projection on the bush which, when in position, passes through a hole in the centre of the top plate.

Bullet support with platform and spring

74. The bullet support is attached to a loose coil spring which, when the magazine is assembled, is positioned outside the spring casing and bears between the underside of the top plate and the bullets, holding these steady in the magazine. To the bullet support is riveted a platform the object of which is to support the last round in the correct position in the lips, ready for engagement by the feed-piece. A stop pin inside the band limits the travel of the platform when the magazine is being loaded.

Loading lever

75. The lever is a flat steel bar which has a hollow bush near the centre and a slot for its retaining pin. A leather carrying strap is riveted to the lever. When the magazine is assembled, the bush fits over the spigot on the magazine body and a hook shaped projection on its outside engages the inner end of the main spring. The correct tension is given to the main spring by rotating the loading lever $3\frac{1}{2}$ times in an anti-clockwise direction and securing it by means of the retaining pin.

Action of the magazine

76. The main spring, through the medium of the top plate, ensures that a round is always in position in the lips of the magazine, ready to be engaged by the feed-piece. As the round is removed, the main spring rotates the top plate which carries with it the ammunition until the next round is retained by the lips.

MAGAZINE NO. 2, MK. I (fig. 15)

77. The No. 2, Mk. I magazine is designed to hold 100 rounds and consists of the following parts :—

(i) Magazine body.
(ii) Follower.
(iii) Top plate with separators, main spring and cover.
(iv) Loading lever with handle and bush.

Magazine body

78. The magazine body consists of a bottom plate and band, spot welded together with catch plates riveted and spot welded at the front and

rear. The catch plates engage the front and rear magazine catches on the gun, when the magazine is in position. A slot is cut in the bottom plate to coincide with the orifice in the gun body and a flange is spot welded to the top of the band. On either side of the slot are lips which project into the orifice in the gun body when the magazine is in position and, being suitably shaped, retain the round until it is pushed out of the magazine by the feed piece. The left lip is movable and pivots on a screw which is spot welded to the bottom plate, a washer, castellated nut and split pin being provided to hold the lip on the screw. The movable lip is normally retained in position by a flat steel spring catch, one end of which is riveted to the outside of the band. On the underside of the bottom plate at the front of the slot is a bullet lead which deflects the round downwards as it is pushed forward.

79. In the centre of the bottom plate is a cylinder which forms an abutment for the noses of the bullets. The cylinder is held in position by a flange on the bottom of a centre post which is riveted to the bottom plate and spot welded. Surrounding the cylinder is a bullet guide of wire which supports and separates the layers of bullets. One end of the bullet guide is spot welded to a bracket which is riveted to the bottom plate, the other end engages a slot in the top edge of the cylinder. In the top of the centre post is a groove for the securing plate of the loading lever and holes for the top plate retainer. A cartridge retaining spring is riveted to the bottom plate near the inner end of the slot to hold the following round while that in the lips is being fed.

80. A carriage head guide ring is welded inside the band and forms a gallery to support the bases of the cartridge cases. Near the lower end of the guide ring and riveted to the bottom plate is a cartridge lead which deflects the cartridges through the slot. Near the top of the guide ring is a stop for the follower.

Follower

81. The follower is placed in the magazine to support the last round in the lips, its shape corresponds approximately with that of a complete round of ammunition. A tail is riveted near the base to keep the follower clear of the feed piece, preventing it from being fed out of the magazine.

Top plate

82. The top plate is assembled on the magazine body and is held by the securing plate of the loading lever which engages the groove and the top plate retainer which engages the holes in the centre post. The separators, which locate the rounds radially in the magazine, are riveted to the underside of the top plate. In the middle of the top plate is formed a shallow recess which houses the main spring. A hollow centre piece is riveted to the underside of the top plate and fits over the centre post to form a bearing for the top plate when the magazine is assembled. Two studs are riveted to the top plate for engagement with the filling machine.

83. The main spring is of clock pattern, its outer end being anchored to a pin which is held in position by a flanged ring riveted to the top plate and spot welded ; its inner end is anchored to a hub having flats to engage corresponding flats on the bush. The spring housing is closed by the top plate cover in which is an inspection window. The window which is

provided with a cover, enables the spring to be examined for tensioning. When the magazine is fully loaded and the spring fully tensioned, no coils should be visible through the window.

Loading lever

84. The lever is a flat steel bar to which is riveted a leather carrying handle. Near the centre of the lever is a hollow bush which fits over the centre post and has flats formed on it to engage those of the hub. Towards one end of the lever is pivoted a securing plate and top plate retainer. The latter also serves to retain the spring tension.

CARE AND CLEANING

85. General instructions for the care, unit storage and cleaning of machine guns are given in A.P. 1641/2 and care of barrels in A.P. 1641/3.

86. Operations to be done before, between and after flights, at the 120-hour inspection of the aeroplane and after 2,500 $\pm$ 500 rounds have been fired are laid down in A.P. 1641B, Vol. II, Part 3, Chapter 2.

87. Certain parts of the gun are permanently assembled and no attempt must be made to dismantle them in the service.

88. When the gun is unloaded the main springs must be eased.

89. Avoid damage to the gun, its accessories and spare parts, by exercising care in handling. As a normal procedure, the gun should be fitted to, or removed from, its mounting as a complete unit. In general, the limited space available in cockpits or turrets renders such operations as the dismantling of the body extension, with the gun in position, very difficult and invariably results in damage to the return springs and rod.

90. Do not clamp the gun in a vice for the purpose of cleaning, inspection or repair. An attachment for securing machine guns to the bench has been introduced and must always be used.

CHAPTER 4

LOADING, UNLOADING, STOPPAGES AND IMMEDIATE ACTION

LOADING AND UNLOADING BY HAND

To load the No. 1, Mk. I Magazine

91. Ensure that there is no tension on the main spring and remove the retaining pin. To ensure that there is no undue frictional resistance, spin the top plate before filling the magazine.

92. Ensure that the platform is visible through the lips and that the hooked end of the loading lever is between 40° and 60° to the left of the front catch plate.

93. Invert the magazine.

94. Hold the magazine with the left hand and take the ammunition in the right hand.

95. Insert a round base first into the lips and push it towards the periphery of the magazine as far as it will go.

96. Rotate the top plate of the magazine, with the fingers of the left hand, until the round is clear of the lips and the next space is visible.

97. Repeat operations 95 and 96 until the magazine is full.

98. Ensure that no space is missed otherwise stoppages may result. If spaces are missed, the extra rotary movement of the magazine during firing will cause an increased impact on the following round and separator, which will tend to distort the separator. As a check upon the loading it is a good plan to have the ammunition arranged in batches of 60 rounds.

99. When the magazine is full the hooked end of the loading lever should coincide with the rear catch plate. A rough check is also provided by noting that, when tensioning the main spring, the loading lever will rotate about $\frac{1}{4}$ to $\frac{1}{3}$ of a turn past the position for the correct number of turns, before maximum tension is reached.

To tension the main spring

100. Place the magazine on the tensioning plate and engage the winding handle with the loading lever of the magazine.

101. Steady the magazine, rotate the loading lever $3\frac{1}{2}$ complete turns in an anti-clockwise direction, replace the retaining pin and remove the winding handle.

To remove the tension from the main spring

102. Place the magazine on the tensioning plate and engage the winding handle with the loading lever of the magazine.

103. Take the weight of the spring and remove the retaining pin.

104. Ease back the winding handle until the tension is removed, and if the magazine is not to be unloaded replace the retaining pin.

To unload the No. 1, Mk. I magazine

105. Remove the tension from the main spring.

106. Remove the top plate and take out the bullet support and platform.

107. Invert the magazine, when the rounds will fall out.

108. Assemble the magazine and replace the retaining pin.

To load the No. 2, Mk. I magazine

109. Ensure that there is no tension on the main spring and disengage the retainer. To ensure that there is no undue frictional resistance, spin the top plate before filling the magazine.

110. Ensure that the securing plate of the loading lever is properly engaged in the groove in the centre post.

111. Ensure that the follower is visible through the lips and that the hooked end of the loading lever coincides approximately with the front catch plate.

112. Invert the magazine and place it on the loading post.

113. Release the movable lip.

114. Hold the magazine with the left hand and take the ammunition in the right hand.

115. Insert a round in the lips and press it into the magazine.

116. Rotate the top plate of the magazine, with the fingers of the left hand, until the round is clear of the lips and the next space is visible.

117. Repeat operations 115 and 116 until the magazine is full.

118. Ensure that no space is missed otherwise stoppages may result. If spaces are missed, the extra rotary movement of the magazine during firing will cause an increased impact on the following round and separator, which will tend to distort the separator. As a check upon loading it is a good plan to have the ammunition arranged in batches of 100 rounds.

119. Replace the movable lip and ensure that it is properly engaged by the spring catch.

To tension the main spring

120. Place the magazine on the tensioning plate and bring the hooked end of the loading lever over the rear catch plate.

121. Engage the winding handle with the loading lever of the magazine.

122. Steady the magazine and rotate the winding handle in a counter-clockwise direction until maximum tension has been applied to the spring. Engage the retainer in the first slot below the position of the maximum tension, taking care to ensure that the retainer is not forced into engagement if it coincides with a slot at maximum tension.

Note.—The No. 2, Mk. I magazine spring must not be wound with any equipment other than the plate and winding handle supplied for the purpose. The equipment supplied for the No. 1, Mk. I magazine is unsuitable for the No. 2, Mk. I magazine.

To remove the tension from the main spring

123. Place the magazine on the tensioning plate and engage the winding handle with the loading lever of the magazine.

124. Take the weight of the spring and disengage the retainer.

125. Ease back the winding handle until the tension is removed, and if the magazine is not to be unloaded, replace the retainer.

To unload the No. 2, Mk. I magazine

126. Remove the tension from the main spring.

127. Release the movable lip.

128. Rotate the top plate when the rounds will fall out.

To load the gun

129. Cock the gun by pulling the cocking handle to the rear as far as it will go, then return the cocking handle to the forward position and ensure that the catch is engaged.

130. Set the safety catch to " SAFE."

131. Hold the magazine with both hands, with the catch plates in the fore and aft line of the gun, and engage the rear catch plate of the magazine with the rear magazine catch. Press down the front of the magazine and engage the front catch plate.

132. Ensure that the magazine is secure by giving an upward pull on the leather handle.

133. The gun is now ready to fire.

To unload the gun

134. Place the palm of the hand against the rear magazine catch lever and grasp the leather handle with the fingers of the same hand. Press forward the lever and lift the magazine, rear end first, from the gun.

135. If the breech block is in the forward position, cock the gun and press the trigger.

136. If the breech block is in the rear position, press the trigger ; cock the gun and again press the trigger.

137. The gun is now unloaded.

STOPPAGES AND IMMEDIATE ACTION

Stoppages

138. The chief stoppages are mainly due to such defects as faulty ammunition, empty or defective magazine, short or broken firing pin, defective extractor or spring, defective feed piece, defective ejector, miss feed and excessive friction. Stoppages can be reduced to a minimum by the careful observance of maintenance instructions.

139. The position of the rear end of the piston, as viewed through the cocking handle slot, or the position of the cocking handle when drawn to the rear to engage the piston, will indicate the action that must be taken to clear a particular stoppage.

Note.—The cocking handle must be drawn back carefully to feel the position of the piston.

The positions are as follows :—

First position.—Rear end of piston below the rear magazine catch lever. Cocking handle fully forward.

Second position.—Rear end of piston to the rear of the rear magazine catch lever. Cocking handle to the front of the rear magazine catch lever.

Third position.—Rear end of piston not visible. Cocking handle below or to the rear of the rear magazine catch lever.

Immediate action

140. Immediate action is the immediate application of a probable remedy for a stoppage, based on the position of the piston and the condition of the gun. It must not be considered to be complete until the gun is again functioning satisfactorily. The immediate action table is set out to give a clear indication of the nature and cause of each stoppage and the probable remedy.

141. The type and number of stoppages which can be cleared in the air will depend upon the installation and the spare parts carried in the aeroplane.

Immediate action table

142.

Position of rear end of piston.	Immediate Action.	Result.	Cause.	Remedy.
First ..	Cock the gun. Fire and watch the top plate of the magazine.	(1) Gun fires.	(1) Miss fire due to defective ammunition.	Nil.
		(2) Gun fires but repeats stoppage.	(2) No feed due to :— (i) Sluggish magazine. (ii) Incomplete backward movement due to (*a*) Friction, dirt or chips. (*Note.*—(ii) (*a*) May also give a stoppage in the second position.) (*b*) Fouling in the gas block or gas plug port.	(2) (i) Change magazine. Continue firing. (ii) (*a*) Cock the gun sharply and fire. If stoppage is repeated, remove magazine, clear gun, wipe away any brass chips which are visible and oil the piston rod through the feed-way in the body. Replace magazine and continue firing. (*b*) Correct maintenance of gun.
		(3) (i) Top plate rotates, but gun does not fire.	(3) (i) Defective firing pin or fouling in the firing pin hole in the breech block.	(3) (i) Remove the magazine. Clear the gun. Change the breech block. Replace the magazine and continue firing.
		(ii) Stopping is repeated, after changing the breech block.	(ii) (*a*) Weak force of blow due to a weak return spring or sluggish action. (*b*) Insufficient protrusion of firing pin.	(ii) (*a*) Correct maintenance of gun. (*b*) Correct gauging and adjustment of firing pin protrusion.

Immediate action table—*continued*

Position of rear end of piston.	Immediate Action.	Result.	Cause.	Remedy.
First ..	Cock the gun. Fire and watch the top plate of the magazine.	(4) Top plate does not rotate and gun does not fire.	(4) No feed due to :— (i) Empty magazine. (ii) Defective or dirty magazine.	(4) (i) Change magazine. (ii) Change magazine. Correct maintenance of magazine.
		(5) Stoppage repeated after changing magazine in (4).	(5) Defective feed piece or feed piece spring.	(5) Remove magazine. Change breech block. Replace magazine and continue firing.
Second ..	Cock the gun. Remove the magazine. **A.** If a round is jammed in the lips, change the magazine and continue firing.	Gun fires.	Defective magazine.	Correct maintenance of magazine.
	B. If a round is correct in the lips. Clear the gun. Empty deflector bag. Replace magazine and continue firing.	(1) Gun fires.	(1) Full deflector bag.	—
		(2) Gun repeats stoppage.	(2) Separated case.	(2) Remove magazine. Clear gun, using clearing plug to remove separated case from the chamber. Replace magazine and continue firing.

Immediate action table—*continued*

Position of rear end of piston.	Immediate Action.	Result.	Cause.	Remedy.
Third ..	Cock the gun. Remove the magazine. **A.** If there is no round in the body or chamber, change the magazine and continue firing.	Gun fires.	Defective magazine.	Correct maintenance of magazine.
	B. If live round only is in body, clear gun. See that round is correctly positioned in magazine lips. Replace magazine and continue firing.	Gun fires.	Point of bullet misses chamber usually because the first round was out of place in the lips before the magazine was put on the gun.	Care in stowage during transport and in fitting to stowage pegs in aeroplane.
	C. If empty case is on face of breech block, clear gun. Change ejector. Replace magazine and continue firing.	Gun fires.	Defective ejector.	—
	D. If empty case is in chamber, clear gun. Replace magazine and continue firing.	(1) Gun fires but repeats stoppage. (2) Gun fires after changing the breech block, but repeats the stoppage.	(1) Defective extractor or extractor spring. (2) (i) Dirty chamber. (ii) Defective or dirty ammunition.	(1) Remove the magazine. Clear the chamber. Change the breech block. Replace the magazine and continue firing. (2) (i) Correct maintenance of gun. (ii) Ensure that ammunition is clean when loading magazine.

Note.—If the gun develops too high a rate of fire, it may cause certain of the stoppages, due to the breech block racing the magazine. The remedy of this is to keep the small gas escape hole in the gas plug clean.

KEY TO ILLUSTRATIONS

Unless otherwise stated the parts are common to both the No. 1 and No. 2 guns.

Part No.	Name of part.	Fig. No.	Part No.	Name of part.	Fig. No.
1	Barrel	1, 3	38	Sear buffer spring sleeve plug with pin.	5, 12
2	Magazine Mk. I ..	1, 14	39	Buffer spring screw with pin.	5, 12
3	Body	1, 4	40	Buffer spring	5, 12
4	Spade grip (No. 1 gun)	1, 4, 6	41	Buffer disc	5
5	Body extension ..	1, 4, 5	42	Buffer	5, 12
6	Gas cylinder	1, 9	43	Safety catch Mk. I (No. 1 gun).	6
7	Shoulder piece (No. 2 gun).	2, 4, 7	44	Safety catch spring plunger (No. 1 gun).	6
8	Trigger guard frame (No. 2 gun).	2, 7	45	Safety catch thumb piece (No. 1 gun).	6
9	Front magazine catch with spring and split pin.	3	46	Safety catch spring plunger guide (No. 1 gun).	6
10	Rear sight bracket ..	3	47	Safety catch spring (No. 1 gun).	6
11	Fore sight bracket ..	3	48	Side piece screws (No. 1 gun).	6
12	Flash eliminator with split pin.	3	49	Spade grip side pieces (No. 1 gun).	6
13	Gas block	3	50	Trigger bar (rear) (No. 2 gun).	7
14	Gas plug with split pin	3	51	Body extension and trigger guard frame securing pin, plunger and spring.	7
15	Body extension and trigger guard frame securing pins.	4, 7	52	Trigger bar (front) (No. 2 gun).	7
16	Deflector bag catch bracket.	4	53	Trigger Mk. II with axis pin and split pin (No. 2 gun).	2, 7
17	Barrel strap	1, 4	54	Safety catch Mk. II (No. 2 gun).	2, 7
18	Spade grip securing screws (2) (No. 1 gun).	4, 6	55	Trigger guard frame bolt with nut and split pin (No. 2 gun).	2, 7
19	Shoulder piece securing screws (2) (No. 2 gun).	4, 7	56	Pistol grip (No. 2 gun)..	7
20	Locking shoulder ..	1, 12	57	Front hand grip (No. 2 gun).	7
21	Deflector bag hinge pin bracket.	4	58	Cocking handle catch with spring.	8
22	Mounting yoke with bolt, washer nut and split pin (No. 1 gun).	1, 4	59	Cocking handle and cocking handle catch pin.	8
23	Top and bottom piston stops.	2, 4, 12	60	Cocking handle with spring.	1, 8
24	Trigger Mk. I with axis pin.	1, 5, 12	61	Cocking handle slide ..	8
25	Trigger spring sleeve cap	5, 12	62	Cocking handle lug ..	8
26	Trigger spring	5	63	Cocking handle lug pin with split pin.	8
27	Trigger spring sleeve ..	5, 12	64	Ejector cover with thumb piece.	1, 8
28	Sear spring	5, 12	65	Ejector	8
29	Sear catch spring ..	5, 12			
30	Sear catch with axis pin	5, 12			
31	Sear with axis pin and guide pin.	5, 12			
32	Sear lever with axis pin	5, 12			
33	Trigger rod with pin ..	5			
34	Trigger rod stop ..	5, 12			
35	Sear buffer spring plunger.	5, 12			
36	Sear buffer spring ..	5, 12			
37	Sear buffer spring sleeve	5			

Part No.	Name of part.	Fig. No.	Part No.	Name of part.	Fig. No.
66	Rear magazine catch lever pin with fixing pin.	8	82	Feed-piece plunger with spring.	11
			83	Extractor	10, 11
67	Rear magazine catch lever.	4, 8	84	Extractor spring ..	10, 11
			85	Firing pin screw ..	10, 11
68	Rear magazine catch with spring.	8	86	Breech block top view	11
			87	Deflector bag frame ..	13
69	Rear magazine catch spring plug with pin.	8	88	Deflector bag hinge pin	13
			89	Deflector bag catch ..	13
70	Piston rod	9, 12	90	Deflector bag	1, 13
71	Hammer..	9	91	Top plate	14
72	Inner return spring ..	9, 12	92	Separators	14
73	Outer return spring ..	9, 12	93	Main spring	14
74	Return springs rod ..	9	94	Loading lever	14
75	Breech block right side view.	2, 10, 12	95	Retaining pin	14
			96	Bullet support with spring and platform.	14
76	Breech block left side view.	10			
			97	Magazine body	14
77	Breech block bottom view.	10	98	Bullet support and guide	14
			99	Lip	14
78	Ejector cam	10	100	Cartridge guide ..	14
79	Firing pin	11	101	Front catch plate ..	14
80	Firing pin spring ..	11	102	Top plate retaining spring clip.	14
81	Feed-piece with axis pin	10, 11			

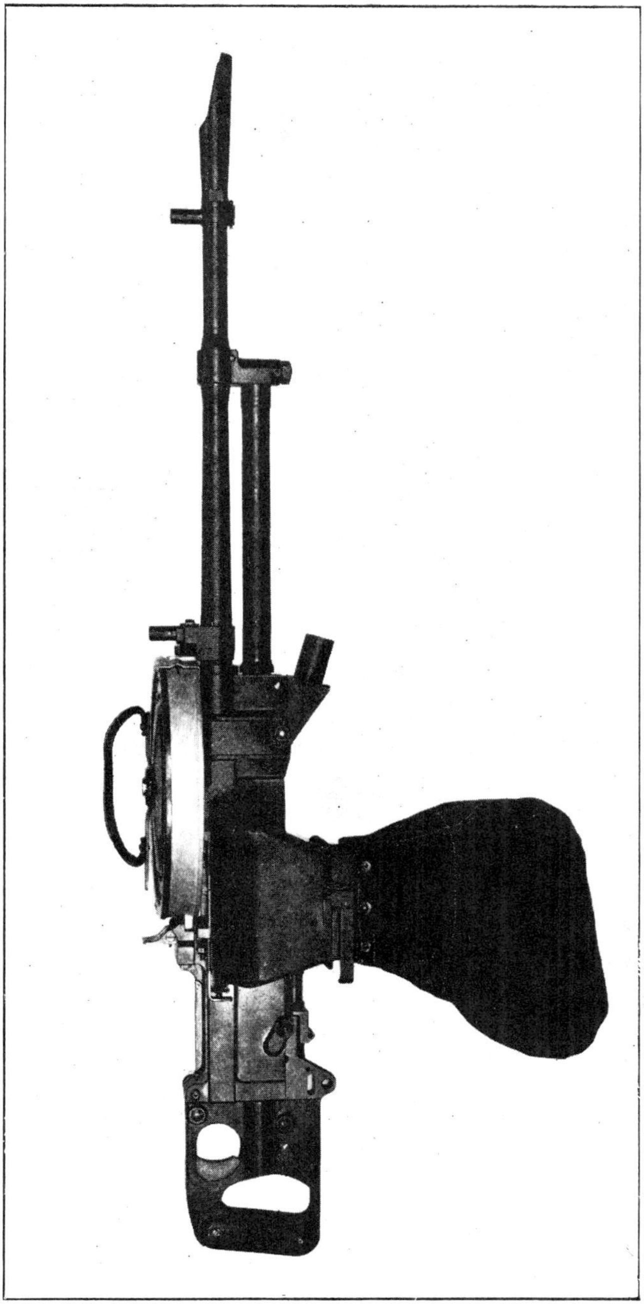

Vickers ·303 in. gas operated Mk. I, No. 1 gun, with No. 1, Mk. I magazine and deflector bag.

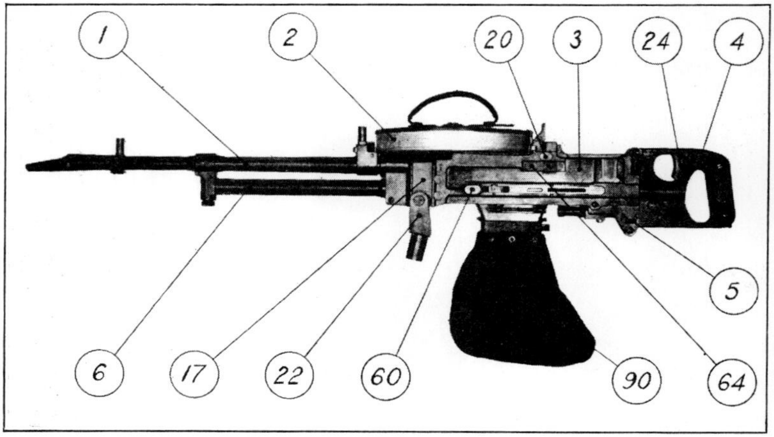

Fig. 1.—Vickers ·303 in. gas operated Mk. I, No. 1 gun, left side view.

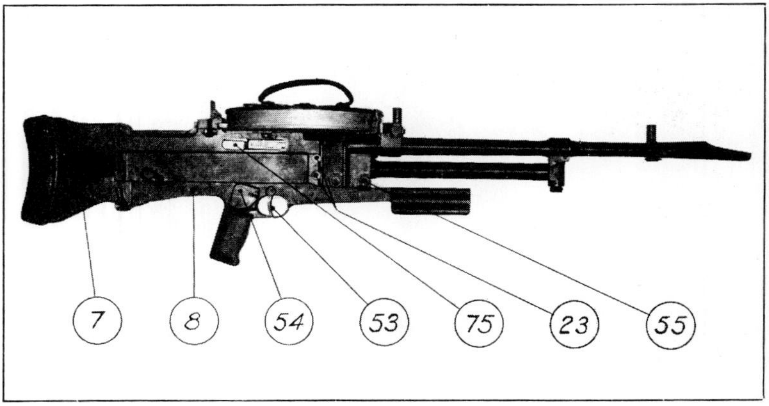

Fig. 2.—Vickers ·303 in. gas operated Mk. I, No. 1 gun, right side view.

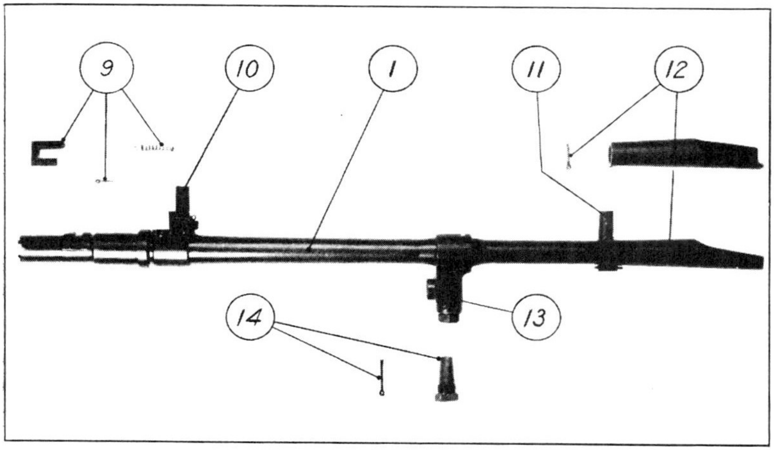

Fig. 3.—Barrel and components.

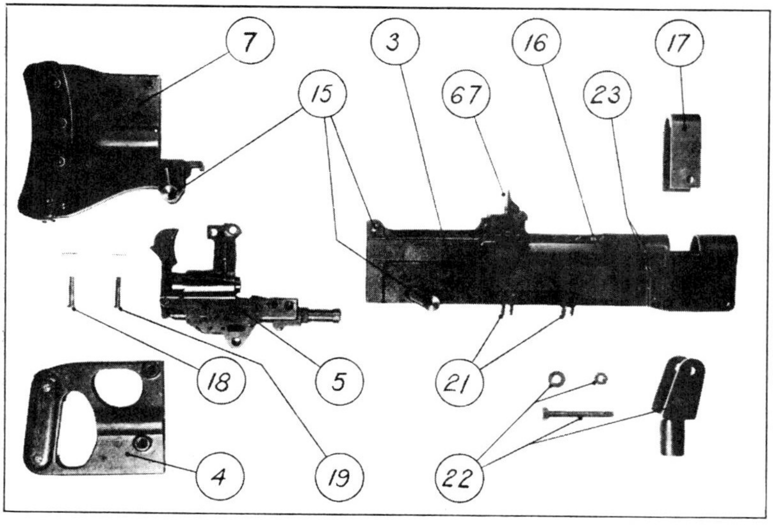

Fig. 4.—Body, body extension and barrel strap. Spade grip and mounting yoke (No. 1 gun). Shoulder piece (No. 2 gun).

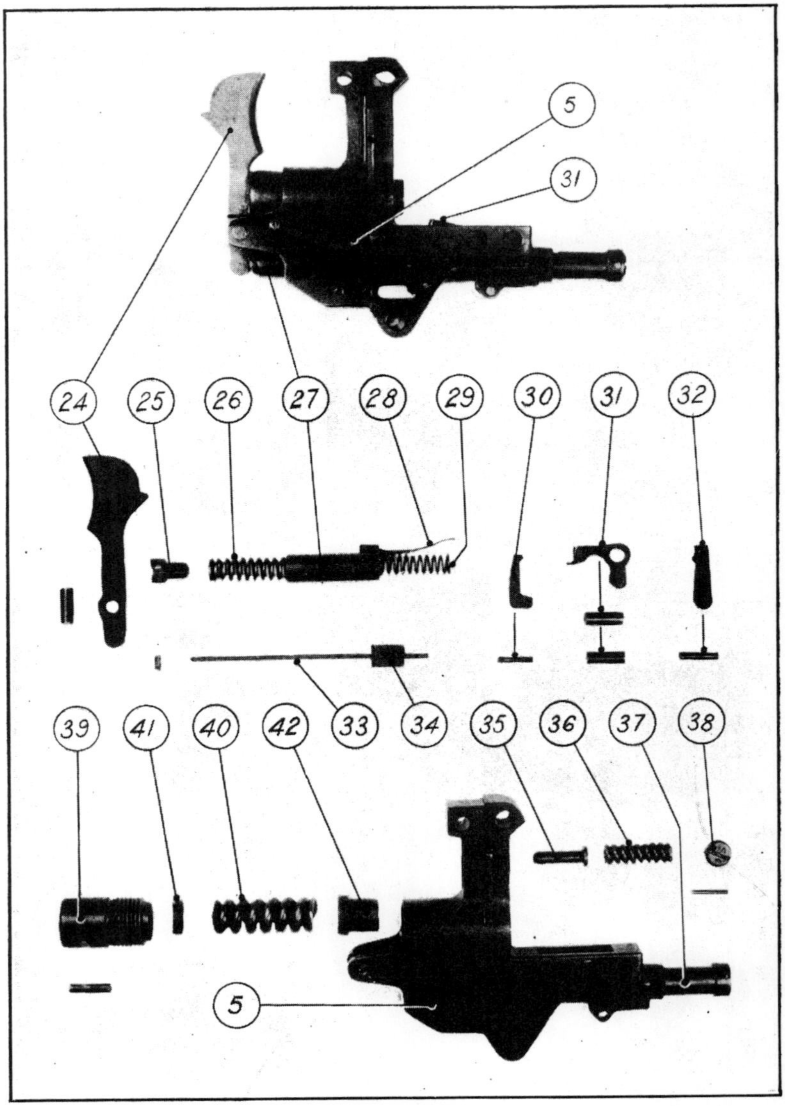

Fig. 5.—Body extension and components.

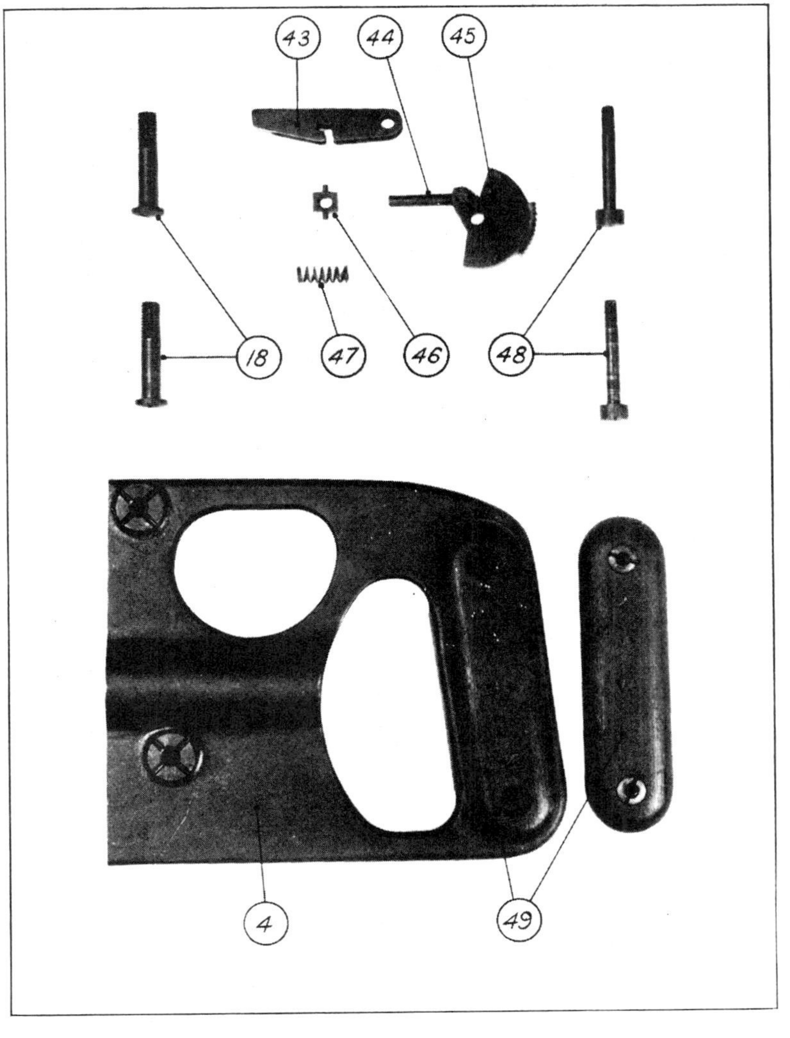

Fig. 6.—Spade grip and components (No. 1 gun).

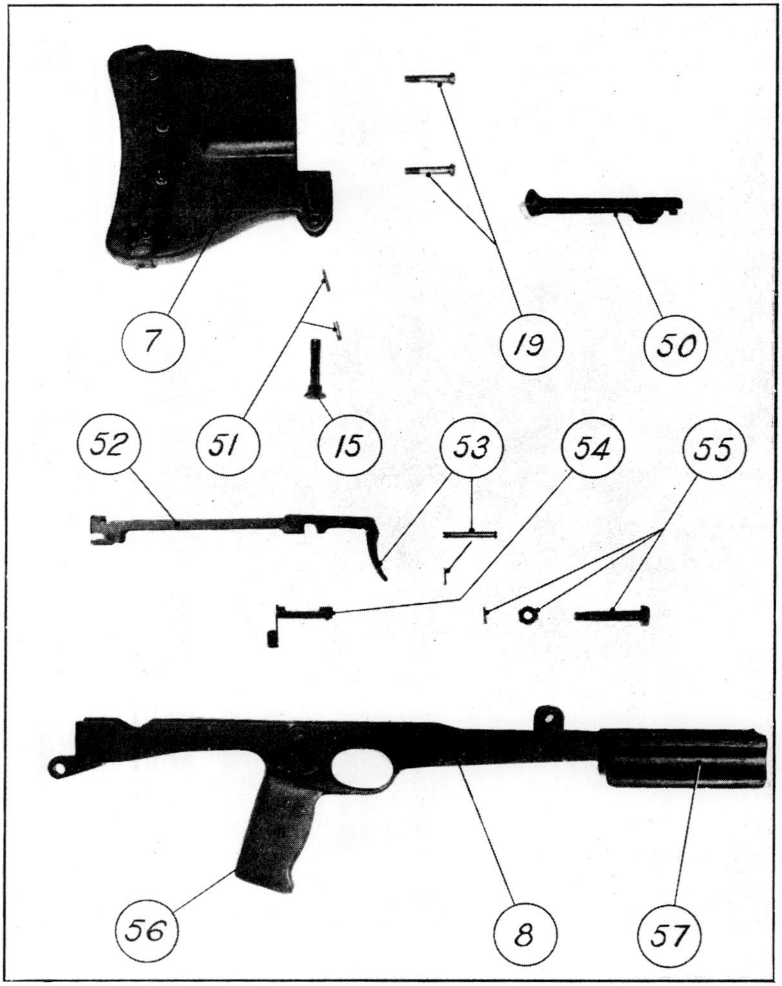

Fig. 7.—Shoulder-piece, trigger guard frame and components (No. 2 gun).

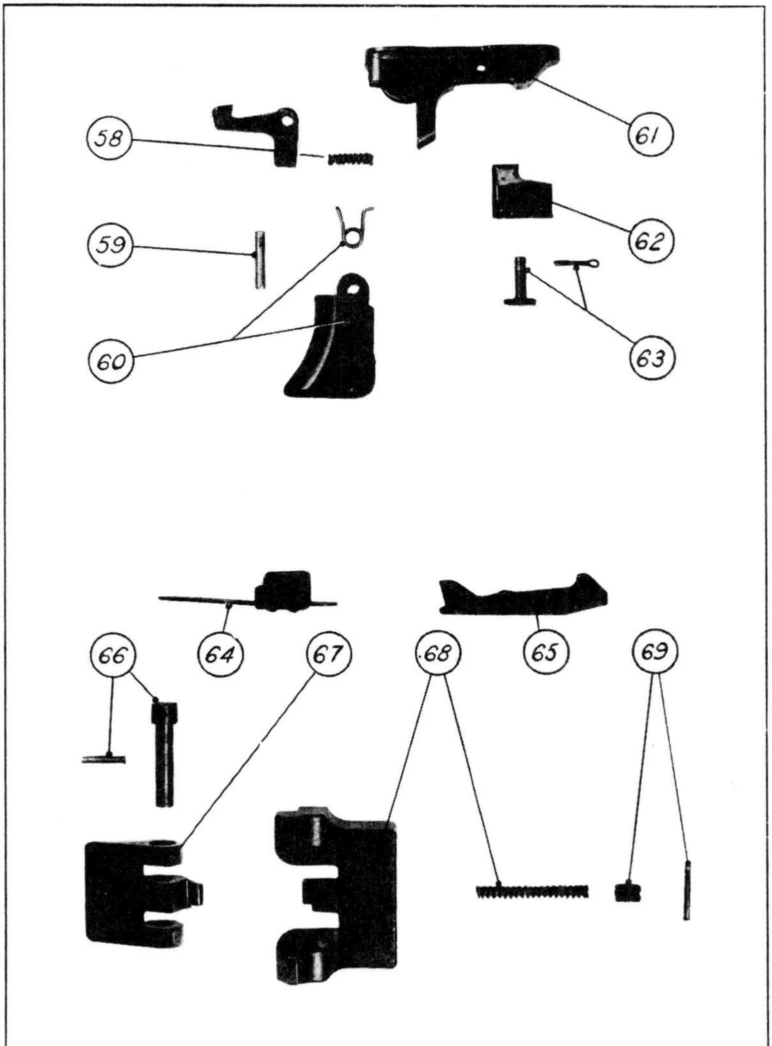

Fig. 8.—Cocking handle components, ejector and rear magazine catch components.

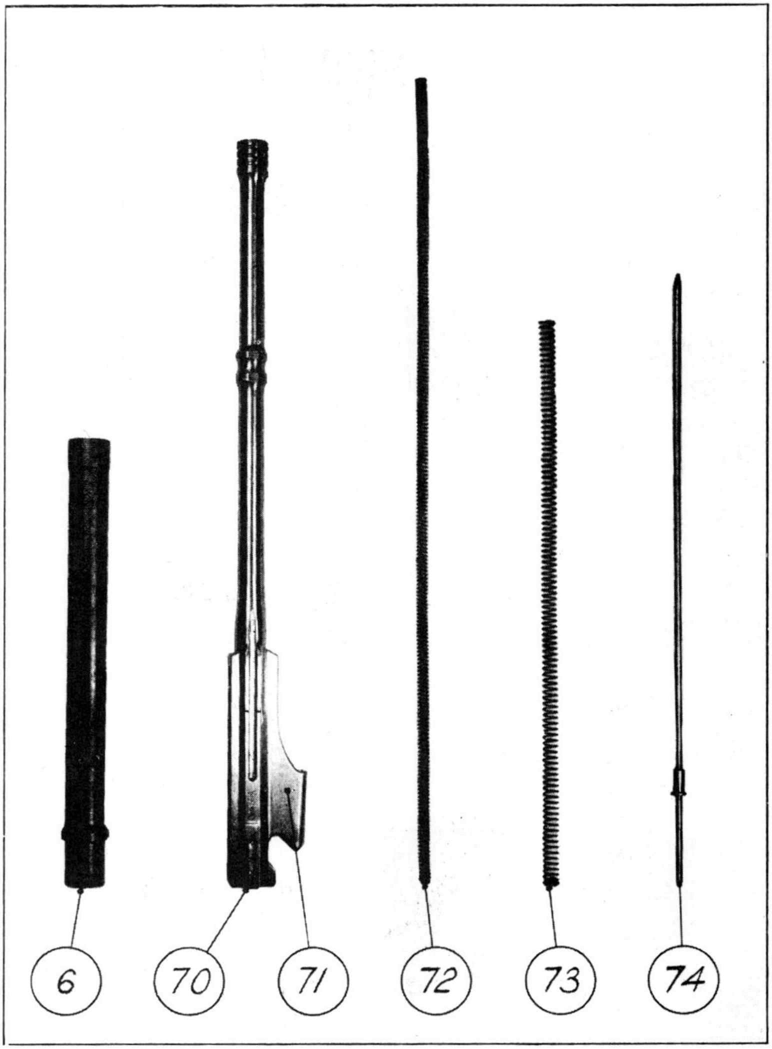

Fig. 9.—Gas cylinder, piston, return springs and return springs rod.

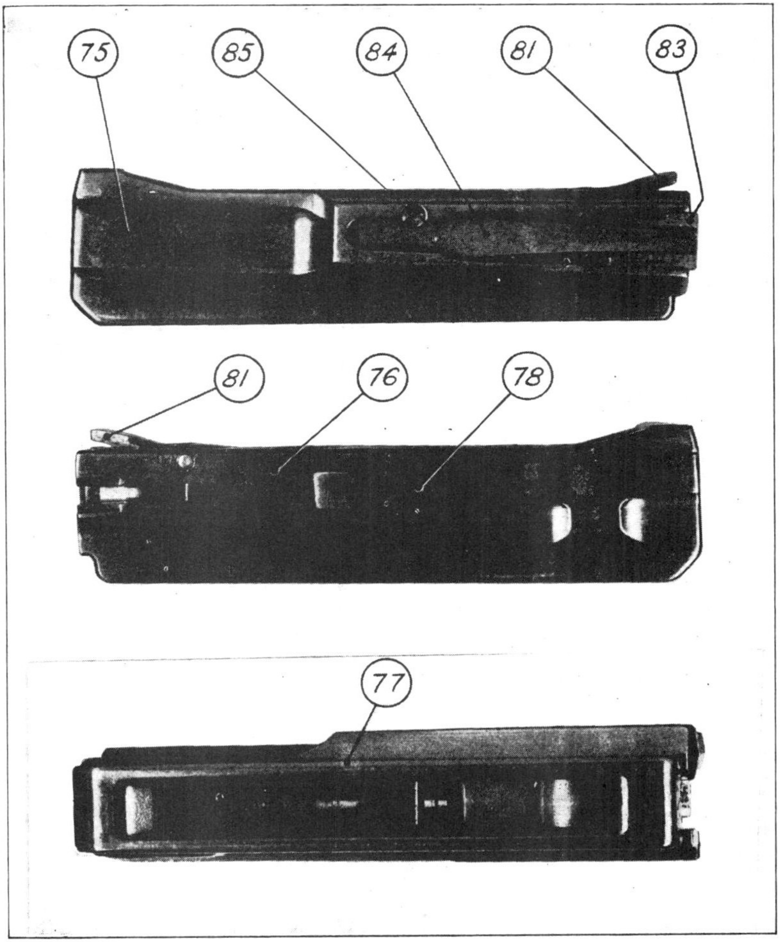

Fig. 10.—Breech block, right, left and bottom views.

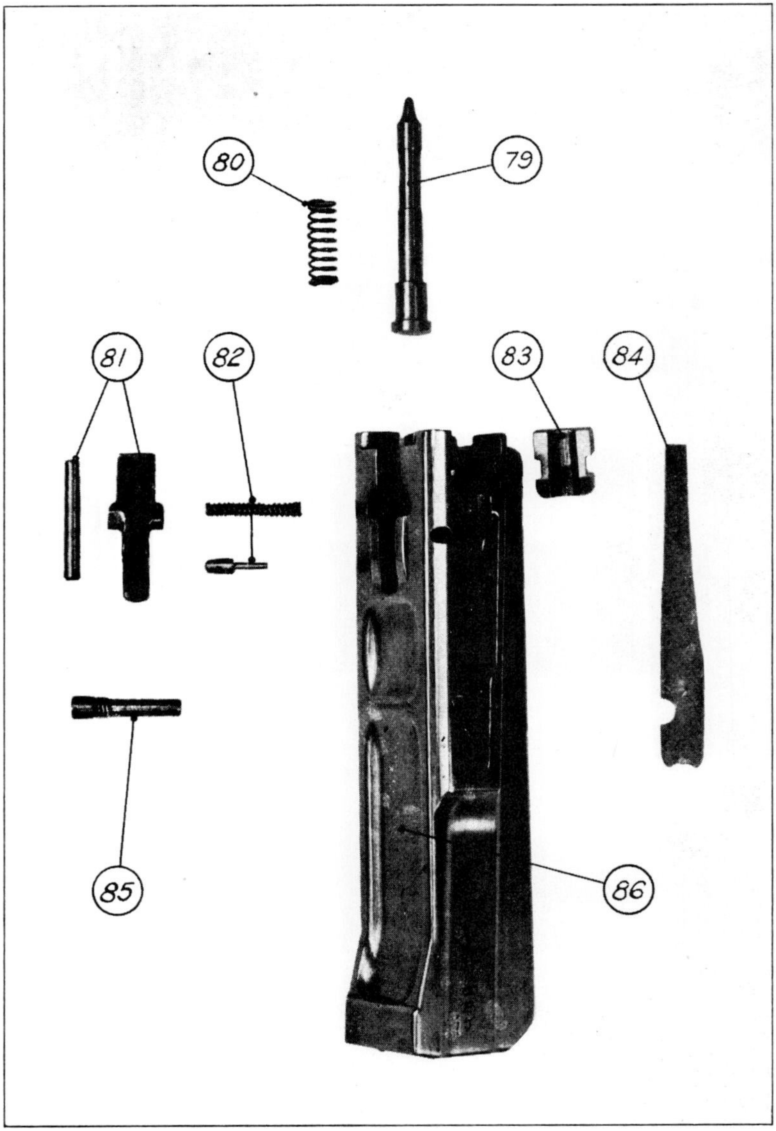

Fig. 11.—Breech block components.

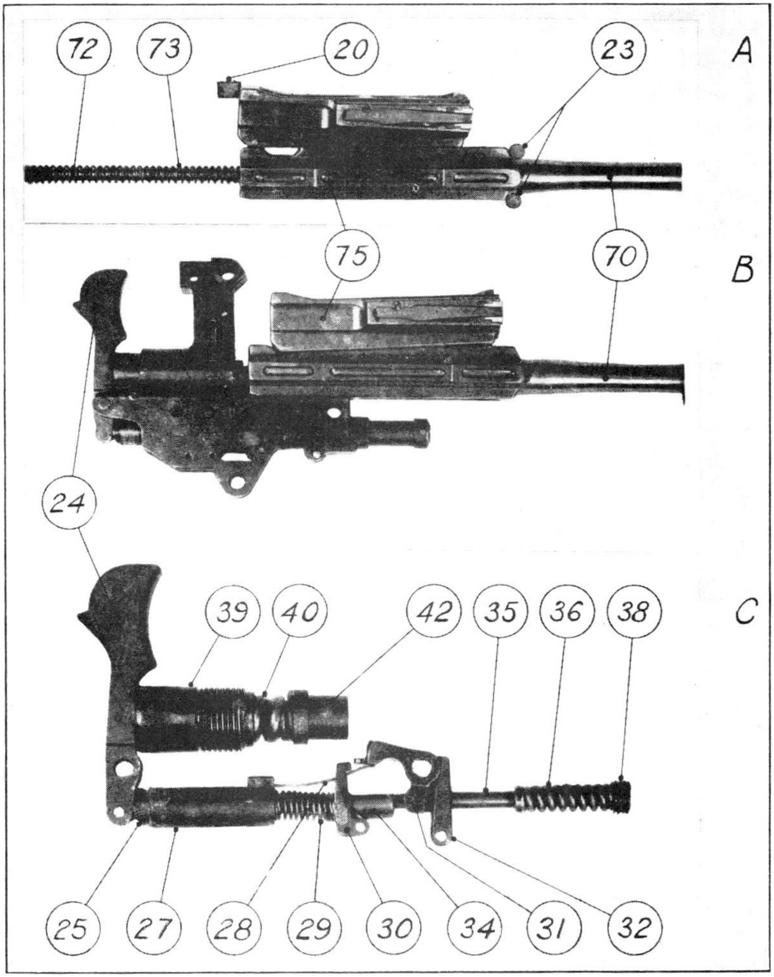

Fig. 12.—Mechanism.
A. Parts forward.
B. Parts back.
C. Trigger mechanism and buffer.

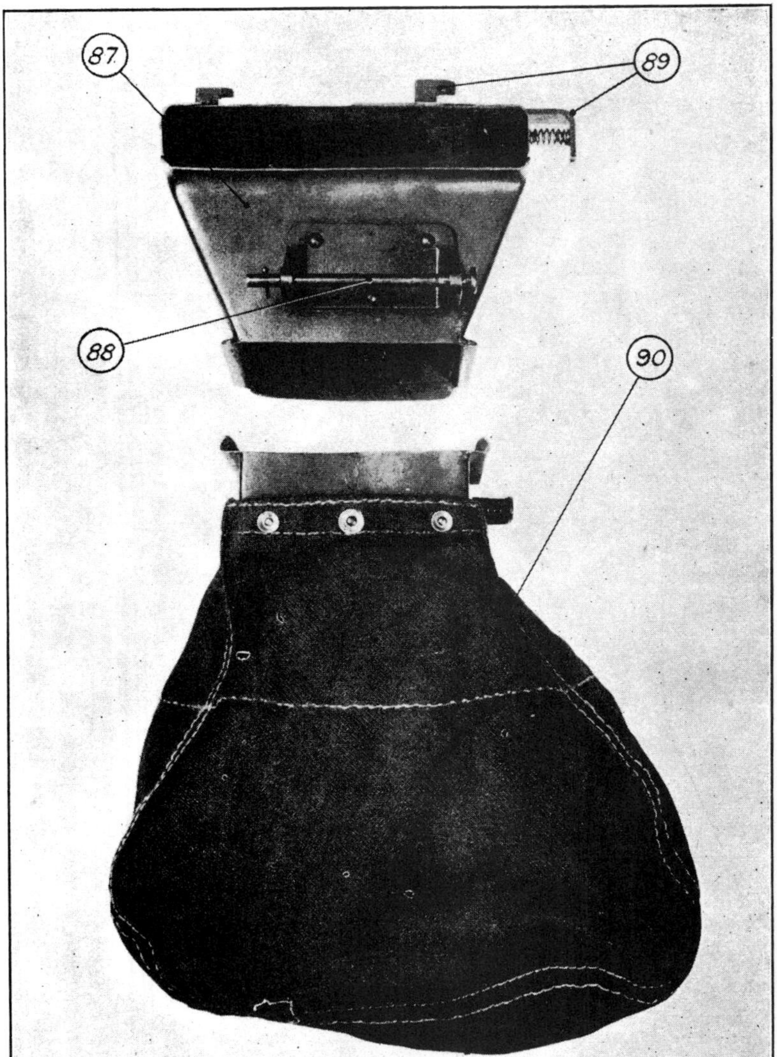

Fig. 13.—Deflector bag.

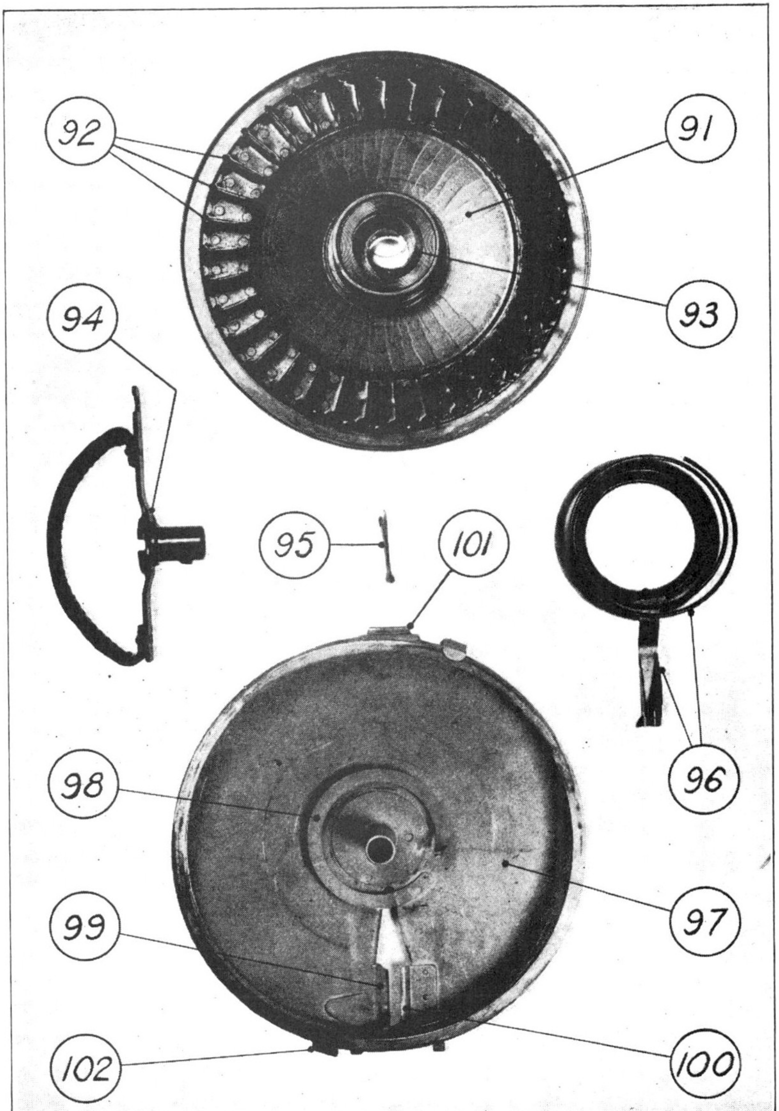

Fig. 14.—Magazine No. 1, Mk. I.

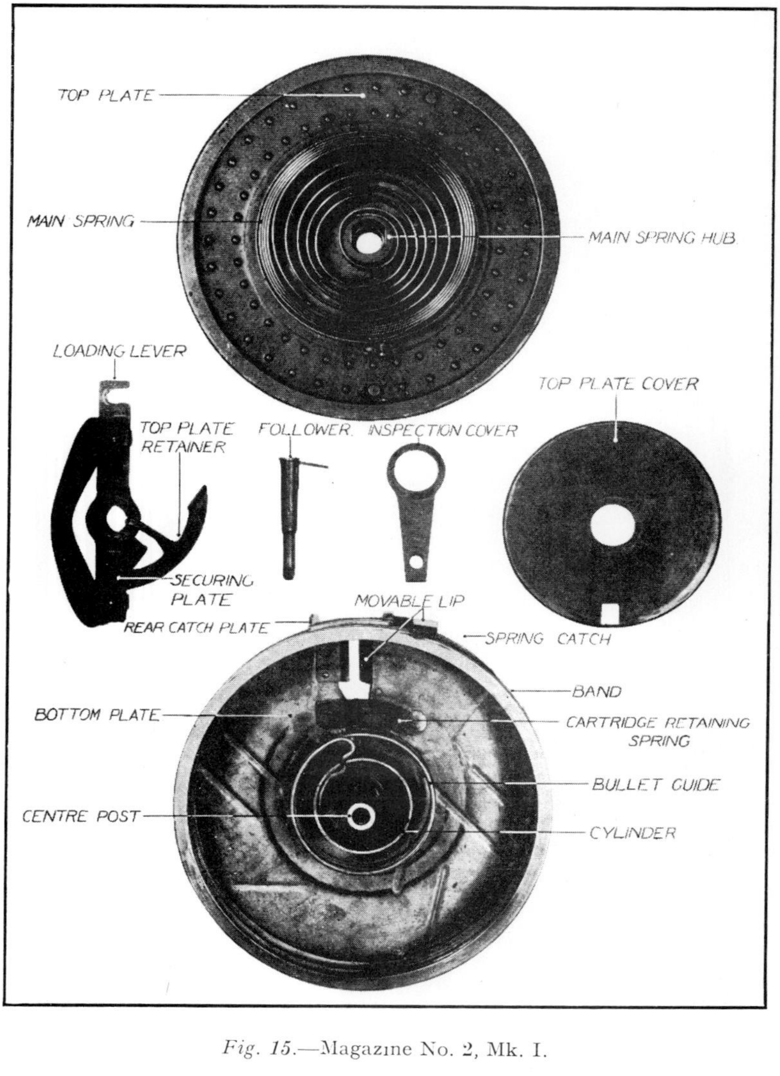

Fig. 15.—Magazine No. 2, Mk. I.

4

Browning .303 in Mark II Gun

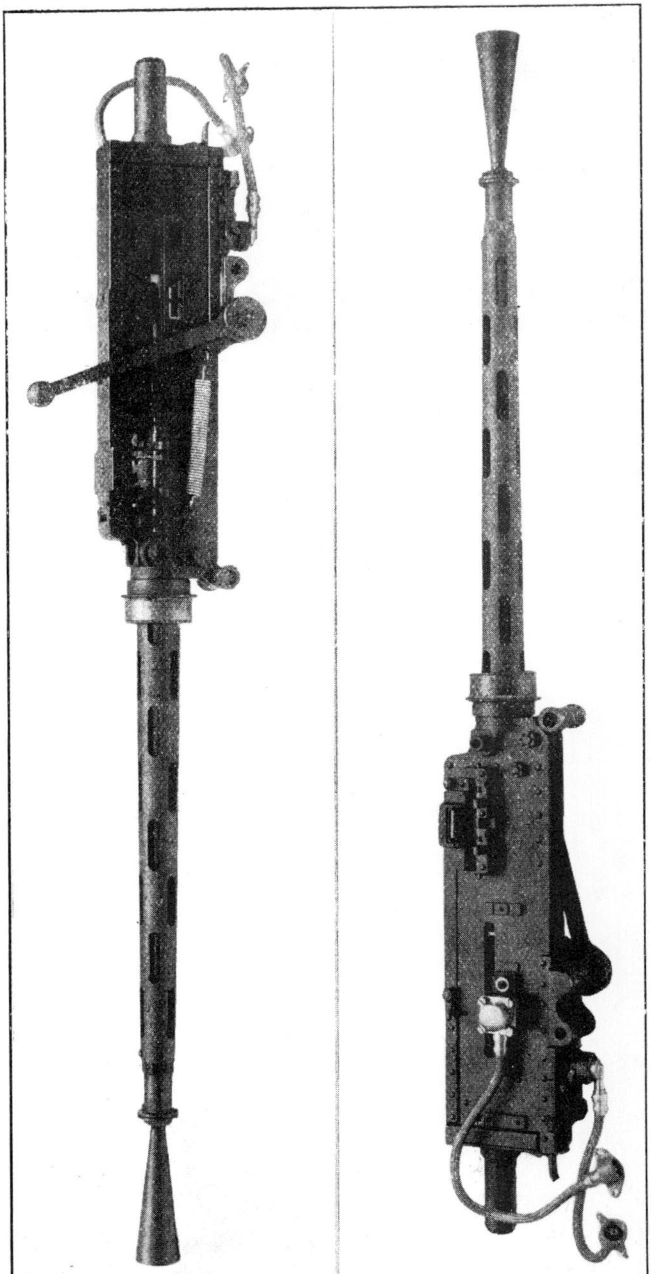

Frontispiece.—Browning ·303 in., Mk. II gun with rear sear release and fire and safe units.

(44282)

CONTENTS

Page

Leading particulars .. 9
List of illustrations .. 10
Introductory .. 11

CHAPTER 1

Description of the gun

Para.

General .. 6

The non-recoiling portions .. 7–30
- Barrel casing .. 8
- Muzzle-attachment .. 9
- Flash eliminator .. 10
- Trunnion block .. 11
- Blast tube adaptor .. 12
- Breech casing .. 13
- Back plate .. 19
- Breech cover .. 20
- Feed lever .. 22
- Feed slide .. 23
- Lock-frame .. 24
- Accelerator .. 25
- Barrel return spring plunger and spring .. 26
- Barrel return spring socket .. 27
- Rear sear .. 28
- Rear sear cradle .. 29
- Rear sear lever .. 30

The recoiling portions .. 31–49
- Barrel .. 32
- Barrel extension .. 33
- Breech block locking-piece .. 34
- Barrel locking spring .. 35
- Breech block .. 36
- Transporter .. 41
- Switch plate .. 42
- Firing mechanism .. 43–48
  - Firing pin .. 43
  - Cocking lever .. 44
  - Sear .. 45
  - Sear retainer .. 46
  - Sear spring retainer .. 47
  - Sear spring retainer keeper .. 48
- Return spring .. 49

CHAPTER 2

Mechanism or action of the gun

General .. 50
Recoil .. 52
Return .. 54
Feed mechanism .. 56
Rear sear mechanism .. 57
Firing mechanism .. 59
Safety arrangements .. 60

CHAPTER 3

Miscellaneous

Loading mechanism 64
Ammunition belt and links 66
Care and cleaning 70

NOTE

This handbook may also be used for the Browning ·303 in., Mk. I, I* and I** guns. These guns only differ from the Browning ·303 in., Mk. II gun in a few details of minor importance and are now obsolescent.

"CHAPTER 4"

"Loading, unloading, stoppages and immediate action."

" Loading and unloading 83–104
To make up the belt by hand 83
Precautions to be observed before loading the gun 87
To load the gun 90
To unload the gun 98

Stoppages and immediate action 105–109
Stoppages 105
Immediate action 107
Immediate action table .. .. .. 109 "

LEADING PARTICULARS

Length of gun, overall, with flash eliminator.	3 ft. $8\frac{1}{2}$ in.
Length of barrel	2 ft.
Weight of gun	21 lb. 14 oz.
Weight of gun with loading mechanism and trigger motor.	22 lb. 12 oz.
Rifling—	
(i) Type	Left hand.
(ii) Twist	1 in 10 in.
(iii) Number of grooves ..	5.
Ammunition used	All types of Mk. VII S.A.A.
Method of feed	Belt.
Speed of fire (approx.)	1,150 rounds per min.
Marks on the gun—	
(i) On the top plate ..	Nature and mark of gun. Maker's initials. Year of manufacture. Number of the gun.
(ii) On the barrel	Proof mark. Calibre. Year of manufacture (last two figures only). Number of the barrel.
(iii) On the breech block. ..	Number of the breech block.
(iv) On various components	Viewers' marks. Broad arrow. Maker's trade mark.
Sights	Separate store.
Mountings—	
(i) Fixed positions ..	Wing installation. Cockpit installation.
(ii) Free positions	Turret installations.

INTRODUCTORY

1. This light machine gun has a high rate of fire and is automatic in action in the sense that so long as the firing control is operated, the gun should continue to fire until all the available ammunition is expended. Ammunition is fed to the gun by belt, which may be of any length according to the stowage space available and subject to the availability of power to feed the belt to the gun. Normally the power for feeding is provided by the gun itself, in which case the effort required must not exceed 11 lb. In some installations, however, there is an external source of power so that belts may be used of a much greater length than would otherwise be possible.

2. Firing of the gun is effected through two units, the rear sear release and the fire and safe, both of which must be operated before the gun can fire. When the gun is mounted for indirect laying in an aircraft, firing is normally effected by remote control through pneumatic, hydraulic or electric mechanisms. In ordinary installations, both the rear sear release and the fire and safe units are operated almost simultaneously by the remote firing button or switch, but in some instances the fire and safe is operated by hand instead of by power; this, of course, is only possible when the gun is within convenient reach of the firer. The gun may also be mounted in aeroplanes to fire between the blades of the airscrew, in which event firing control is through the ordinary synchronising gear which in turn may be controlled by one of the above mentioned power methods. When synchronising gear control is employed, one unit on the gun (the rear sear release) is operated pneumatically, hydraulically or electrically, the second unit (the fire and safe) is replaced by the synchronising gear trigger-motor. Whichever method of control is used, it is self-sufficient, self-contained and exclusive.

3. The gun depends for its action on the recoil of internal moving parts assisted, as the bullet leaves the barrel, by gas pressure. After the first shot the recoil of these parts ejects the empty case and the return reloads the gun and also supplies the power to draw the cartridge belt into position for feeding, which may be from the right or left side of the gun. The belt is guided to the gun by a chute or rollers and, after firing, the empty cases and separated links are discharged from the gun.

4. Cooling of the gun is entirely by air flow over the barrel, muzzle-attachment and flash eliminator, which are the only parts that need cooling, but heating may be provided to prevent freezing of the working parts.

5. The gun may be installed in one of three positions in aeroplanes :—

(i) Wing installation.

(ii) Cockpit installation.

(iii) Turret installation.

The first two installations give fixed gun positions, the third gives a free gun position. In positions (i) and (iii), firing control is by one of the power methods mentioned above except that in some turrets, hand instead of power operation is provided for the fire and safe unit. In position (ii) firing is between the blades of the airscrew and is controlled through the synchronising gear.

CHAPTER 1

DESCRIPTION OF THE GUN

General

6. The gun may be divided naturally into two parts for purposes of description :—

(i) The non-recoiling portions.
(ii) The recoiling portions.

THE NON-RECOILING PORTIONS

7. The non-recoiling portions are those parts of the gun that are visible externally and also one internal unit. The external visible parts are :—

(i) The barrel casing, which includes the muzzle-attachment and flash eliminator.

(ii) The trunnion block.

(iii) The blast tube adaptor.

(iv) The breech casing, which includes the right and left side plates with their incorporated fittings, the top and bottom plates, the back plate, the breech cover, and feed mechanism.

The internal unit is :—

(v) The lock-frame, which houses the accelerator. barrel return spring, rear sear and their components.

Barrel casing (fig. 1)

8. The barrel casing is a tapered drawn steel tube with slots cut in it to allow air access to the barrel for cooling purposes. It is screwed and sweated into the trunnion block at its rear end and into its front end is screwed and secured by a set screw the muzzle-attachment. The function of the barrel casing is to hold and to act as a guide for the barrel when this recoils.

Muzzle-attachment (fig. 1)

9. The muzzle-attachment is a hollow steel plug open at both ends ; the rear end forms the barrel bearing and the front end is threaded to take the flash eliminator. Externally at the rear a hexagon is formed and at the front there is a flange with a number of holes to enable the flash eliminator to be secured by its split pin.

Flash eliminator (fig. 1)

10. The flash eliminator is a conical tube having an internal angle of 15°. It is screwed into, and has a similar flange to, the muzzle-attachment for the split pin. The small end of the flash eliminator is coned and forms a choke for the front

end of the muzzle-attachment. As the bullet leaves the muzzle the escaping gas impinges on this cone which is chromium plated to prevent the accumulation of fouling. Care must be exercised in the cleaning of this portion of the eliminator to prevent damage to the plating. It is important that the whole of the flash eliminator be kept clean as an excess of fouling would adversely effect the accuracy of the gun and cause stoppages. The flash eliminator must always be screwed up tightly and secured by its split pin, otherwise it may come off during firing with risk of injury to the firer.

Trunnion block (figs. 1, 4 and 12)

11. The trunnion block is cylindrical in its front portion which is threaded internally to take the rear of the barrel casing and externally to take the blast tube adaptor. Above and to the rear of the cylindrical portion, two wings form a gap to receive the hinge of the breech cover. Between the wings is riveted a bracket which houses the breech cover detent. Extending to the rear in line with the axis of the barrel casing the upper surface of the trunnion block forms the bottom of the feed opening. The trunnion block is bored through to receive the breech end of the barrel of which it thus forms the rear bearing. Projecting downwards are two flanges to which the side plates of the breech casing are riveted. These flanges also form the support of the forward mounting bracket which is secured by two bolts and nuts.

Blast tube adaptor (figs. 1 and 4)

12. The blast tube adaptor consists of a pair of circular flanges resembling a cotton reel in appearance under which name it is sometimes described colloquially. It is located at the extreme rear end of the barrel casing and its function is to close the tube in the wing or other housing in the aeroplane for the barrel portion of the gun, the space between the two flanges being packed with asbestos and bound with a brass strip.

Breech casing (figs. 1 and 4)

13. The breech casing constitutes a housing for all the essential working parts of the gun with the exception of the barrel. Thus the gun may be regarded externally as two casings which are connected by the trunnion block. The breech casing is a rectangular box made up of four steel plates riveted together and a removable back plate, the two side plates being connected by the top and bottom plates neither of which extends the whole length of the side plates. The forward end at the top is closed by the breech cover.

Side plates

14. The side plates extend from the trunnion block to the rear end of the gun, being cut away at the front to form the feed opening. A bracket is riveted to the front of each plate and level with the bottom of the feed opening. These brackets are suitably machined to accommodate the front and rear cartridge stops, the bullet stop, the retaining pawl complete with spring, the filling piece and the four ammunition-box neck lugs, which fittings are held in position by retaining pins secured by split pins. The location of these fittings depends upon the direction of feed.

15. In each side plate is cut a slot through the lengths of which projects the cocking stud. This stud is engaged by the loading lever when the loading handle is operated for the initial loading of the gun. As the gun may have either right or left hand feed exactly similar slots are cut in each side plate and the cocking stud may be inserted from either side, the same applying to the fire and safe unit or trigger-motor which fits into a short rectangular slot underneath the cocking stud slot. On the inside of each plate and parallel to the bore of the barrel are riveted two breech block guides which control the breech block when it is in the rear position. Near the top and immediately to the rear of the feed opening on the inside of the left plate only are the transporter front and rear cams, each secured by rivets, the functions of which are explained in the description of the action of the transporter in Chapter 2, para. 53 and 54.

16. On the inside of each side plate near the bottom is a groove extending forward from the rear end for a distance of about 3 in. These grooves position the lock-frame ; there is a small hole in the left groove into which enters a spring-loaded retainer plunger on the lock-frame which holds the latter in position when the back plate is removed. In the rear of the right side plate is a slot to accommodate the stud of the return spring rod. At the rear end of each plate is a vertical projection down which the back plate slides to close the rear of the breech casing.

Bottom plate

17. The bottom plate is flanged and extends for about one-third of the length of the breech casing from the rear, the remainder of the space between the side plates being left open for the ejection of the empty cartridge cases. At its forward end the bottom plate carries the locking-piece cam which is secured by a nut and split pin. Two lugs are provided for the attachment of the gun to the mounting and two others for the attachment of the loading mechanism. Interrupted projections are formed on the sides of the bottom plate for the

attachment of the rear sear release unit, and a slot is cut in the plate to allow the rear sear lever to be engaged by the plunger of the rear sear release unit.

Top plate

18. The top plate is riveted to the side plates towards the rear and on its under-side has a bracket which actuates the cocking lever as the breech block slides past. The top plate extends for about one-third of the total length of the breech casing, the remainder of the top of the box being closed by the breech cover which is essentially a separate continuation of the top plate and is hinged at its forward end.

Back plate (figs. 1 and 2)

19. On either side of the back plate are dovetailed grooves which, engaging the corresponding projections on the side plates, hold the back plate in position to close the rear end of the breech casing and form a bearing for the lock-frame and return spring rod. At the bottom is a thumb-operated catch which secures the back plate to the breech casing. A tube which is screwed to the rear of the back plate houses a buffer and twenty-two fibre discs which take the shock of impact at the end of recoil. The tube is closed by a screwed plug which is secured by a spring loaded plunger.

Breech cover (figs. 1, 4 and 5)

20. The breech cover, which has to be raised for initial loading of the gun and as a preliminary to any dismantling, closes the top front portion of the breech casing under all normal circumstances. It is hinged at its forward end to the trunnion block and held at the rear by a thumb-operated catch which engages the top plate of the breech casing. The hinge has serrations into which bears a spring-loaded detent so that the cover may be lowered only in stages. It should be knocked by hand to carry it past each catching stage, the purpose of which is to prevent the cover from banging down if it should be opened for any reason in an air stream.

21. On the inside of the breech cover is riveted the feed lever stud and the transporter ramp. This ramp, in addition to guiding the transporter, serves as the rear anchorage of the transporter guide spring, which is held at its forward end by means of a stud. Towards its forward end the breech cover is suitably shaped and grooved to accommodate the feed slide.

Note.—The breech cover must not be lowered unless the breech block is in its forward position, otherwise the feed lever will not engage the correct groove in the breech block and the subsequent closing of the breech would damage the feed lever and breech block.

Feed lever (figs. 4 and 5)

22. The feed lever is held in position on its stud in the breech cover by a bush and split pin. Two holes are drilled in the lever near its axis and into one of these holes, depending upon the direction of feed, is assembled a plunger and spring. The plunger retains the lever in the correct position for engaging the appropriate groove in the breech block, providing this is closed, when the cover is lowered. At one end of the lever is a stud for engagement with the groove, while at its other end the lever is shaped to engage a recess in the feed slide.

Feed slide (figs. 4 and 5)

23. The feed slide has a guide on either side to engage the grooves in the breech cover and carries the feed pawl. A lateral movement is transmitted to the feed slide as the feed lever oscillates about its stud under the influence of the moving breech block, causing the feed pawl to feed in the belt. The feed pawl is kept up to its work by a spring and has a leg attached to prevent double feed.

Lock-frame (figs. 6 and 7)

24. The lock-frame is a rigid assembly of two steel plates, a separator and a rear sear stop riveted together, the latter acting as a rear separator. Each plate is rectangular in shape and has (1) a prong at its forward end which is bevelled to engage the pin of the breech block locking piece, (2) a gap halfway along the top to accommodate projections on the rear sear cradle, (3) a rib on the outside to act as guides for inserting and holding the lock-frame in position in the breech casing and (4) suitable holes drilled for the pins of the various components. A stud is riveted to the right side plate to form a stop for the accelerator in its rearmost position and in the left side plate a slot is cut to provide a guide for the barrel return spring plunger. The separator is bored to receive the barrel return spring socket and the rear sear stop is cut away to clear the rear sear, a projection on the stop forming a bearing for the rear sear plunger. Situated at the rear of the lock-frame is the lock-frame retainer. This consists of a spring-loaded plunger and stud housed in a tube, the plunger and stud projecting through holes in the lock-frame to hold the retainer in position.

Accelerator (figs. 6, 7 and 12)

25. The accelerator is hinged at the forward end of the lock-frame and has two curved horns each with a ramp on the rear concave face. The accelerator assists the rearward movement of the breech block and locks the barrel extension and lock-frame together, through the engagement of the ramps with a T-shaped projection on the barrel extension, until the breech block is returned to the firing position.

Barrel return spring plunger and spring (figs. 6 and 7)

26. The barrel return spring plunger passes through the barrel return spring which bears against the inside of the barrel return spring socket. The head of the plunger is shaped to bear against a T-shaped projection on the barrel extension and has a guide pin which projects into a slot in the left side plate of the lock-frame.

Barrel return spring socket (fig. 7)

27. The barrel return spring socket is cylindrical in shape and is bored to accommodate the barrel return spring and plunger. It has a head formed at its forward end through two grooves in which pass the securing pins to hold it in position in the separator. The rear of the head forms an abutment for the rear sear buffer spring.

Rear sear (figs. 6, 7 and 12)

28. The rear sear is hinged to a cradle and has a wide bent across its upper surface which engages a projection or bent on the breech block to hold this in the open position when firing is stopped. At the rear of the sear are two projections on which the prongs of the rear sear lever operate. A spring-loaded plunger which bears against a projection on the rear sear stop raises the rear sear when the lever is released.

Rear sear cradle (figs. 6, 7 and 12)

29. The rear sear cradle pivots on suitable bearings riveted on the inside of the lock-frame plates. Its front face is machined to accommodate the rear sear buffer spring washer and a projection on each side, extending into the gaps in the lock-frame plates, limits its movement. A hole through the centre of the rear sear cradle provides a clearance for the barrel return spring socket upon which is situated the rear sear buffer spring.

Rear sear lever (figs. 7 and 12)

30. The rear sear lever is pivoted at the rear of the lock-frame. Its front end projects below the base of the lock-frame where it is engaged by the plunger of the rear sear release unit. At the rear the lever is forked to engage the projections on the rear sear to lower this when the release unit is operated.

THE RECOILING PORTIONS

31. The recoiling portions are those internal parts that are moved to the rear by the shock of discharge and are returned to the firing position by the energy conserved during recoil.

They consist of :—

(i) The barrel.

(ii) The barrel extension with breech block locking-piece and barrel locking spring.

(iii) The breech block which carries the transporter, switch plate and firing mechanism.

(iv) The return spring.

The action of each of these parts in the working of the gun is described in Chapter 2.

Barrel

32. The barrel is bored longitudinally and rifled to suit S.A. ammunition. At the rear the bore is increased in diameter to form a chamber which is coned to conform to the contour of the Mk. VII cartridge case. Between the chamber and the commencement of the rifling is a second cone called the lead. Externally the barrel is parallel for a few inches at each end to provide bearing surfaces in the muzzle attachment and trunnion block. The muzzle bearing is chromium plated. To the rear of the breech bearing the barrel is threaded and serrated for engagement with the barrel extension and locking spring, the thread enabling the cartridge head space of the gun to be adjusted. Between the parallel portions the barrel tapers towards the muzzle.

Barrel extension (figs. 8 and 12)

33. The barrel extension is machined from solid steel and is cut away in the centre for practically the whole of its length. The front end is threaded internally to engage the threaded portion of the barrel, externally it is cut away to clear the transporter cam and the ejector and in the left side is a groove with undercut for the barrel locking spring. Cut along the full length of the inside are two grooves which engage guides on the breech block. A vertical slot is cut in the rear to accommodate the breech block locking-piece and to allow vertical movement elongated holes are provided for the locking-piece pin. Longitudinal slots are cut on the outside to accommodate the prongs of the lock frame and so allow the prongs to disengage the locking-piece from the breech block on recoil. Clearances are also cut in the sides at the rear to allow room for the plunger of the trigger-motor or fire and safe unit, whichever is fitted. Two grooves are cut in the underside to engage ribs on the locking-piece cam and so check any tendency of the barrel extension to lift when the locking piece rides up the cam at the end of the forward movement. At the rear is a T-shaped projection with stud which engages the barrel return spring plunger.

Breech block locking-piece (fig. 8)

34. The breech block locking-piece, which locks the breech block to the barrel extension in the firing position, is a rectangular steel block having a transverse hole drilled

through it for the locking-piece pin. The top of the front face is chamfered to provide a lead into the locking recess in the breech block in the event of the breech block not being fully home when the locking-piece engages the cam in the breech casing.

Barrel locking spring (figs. 8 and 12)

35. The barrel locking spring is a flat steel spring having a projection at its forward end to engage the notches at the breech end of the barrel, when this is assembled in the barrel extension.

Breech block (figs. 9 and 10)

36. The breech block is a rectangular steel block which is lightened as far as possible by the removal of unnecessary metal. The front face is prepared with undercut guides, the tops of which are chamfered, to engage the rim of the cartridge case. Towards the bottom of this face a hole is drilled for the firing pin.

37. Two interrupted cam grooves are cut in the top of the breech block for engagement with the feed lever stud. The appropriate grooves for right or left hand feed are completed by the switch plate which is accommodated in a circular recess in the top of the breech block. A hole is drilled in the centre of this recess for the axis of the switch plate and a second smaller hole to one side for the switch plate retaining plunger. The letters R and L indicate the correct position of the switch plate for right or left hand feed. Behind the switch plate recess a slot is cut for the cocking lever and the plate of the sear spring retainer keeper, a vertical hole being drilled through the breech block for the stem of the latter. Towards the rear of the slot are grooves to facilitate the positioning of the sear spring retainer.

38. The rear of the breech block is recessed and grooved to accommodate the sear, sear retainer and sear spring retainer. Two holes are drilled for the firing pin and the return spring respectively.

39. On each side of the breech block there is a guide for engagement with the grooves in the barrel extension. A hole is drilled horizontally through the breech block for the cocking stud which can be assembled from either side according to the direction of feed, pins being provided to ensure correct positioning of the cocking stud. A second hole is drilled near the rear for the cocking lever axis pin. Towards the front of the block and in the left side a hole is drilled for the transporter axis; this hole breaks through the axis hole for the switch plate so enabling the transporter to hold the switch plate

in position, an undercut being provided to hold the transporter. Also towards the front of the breech block, and on the left side, is a stop which limits the downward movement of the transporter.

40. On the underside of the breech block at the rear, is a projection or bent with which the rear sear engages to hold the breech block in the open position. This projection is also engaged by the horns of the accelerator, when the gun recoils, to throw the breech block to the rear, the reverse action taking place when the breech block moves forward. In front of this projection there is a recess which takes the locking piece when the breech block is in the forward, i.e. firing position. The breech block is thus locked to the barrel extension and barrel to take the shock of discharge.

Transporter (figs. 9, 10, 11 and 12)

41. The transporter is in the form of a lever with an axis pin at one end and a projection at the other. A radial flange near the axis pin engaging the undercut in the breech block holds the transporter in position in the breech block. The projection is drilled from the outside to house a spring loaded plunger which is retained by a pin and raises the transporter, by engaging the front cam in the breech casing, as the breech block travels forward. The underside of the projection has a claw formed on it to engage the rim of the cartridge case and is slotted to accommodate the ejector with its spring, the ejector being retained by a pin. On the top the transporter is suitably shaped to engage the transporter ramp and spring in the breech cover.

Switch plate (figs. 9, 10 and 11)

42. The switch plate is circular in shape, being relieved on each side to conform to the contour of the breech block. A groove is cut in the top to complete the cam grooves in the breech block while its axis is provided with a circumferential groove in which fits the axis of the transporter to hold the switch plate in position. Two small holes are drilled in the switch plate for the head of the spring-loaded plunger which is housed in the breech block and retains the switch plate in the correct position for the required direction of feed.

Firing pin (figs. 9 and 11)

43. The firing pin is cylindrical in shape, being drilled from the front to accommodate the spring, which is retained in position by a pin, and has a bent formed on either side at the rear. A slot is cut in the firing pin to accommodate the cocking lever, also to enable the stem of the sear spring retainer keeper to pass through it and form a rear bearing for the spring.

Cocking lever (figs. 9, 10, 11 and 12)

44. The upper end of the cocking lever is shaped to engage the cocking lever bracket and its lower end is radiused to bear against the rear of the slot in the firing pin. The cocking lever pivots on its axis pin as the breech block travels to and fro and withdraws the firing pin to the cocked position during the rearward movement of the breech block.

Sear (figs. 9, 10 and 11)

45. The sear is accommodated in a horizontal recess in the rear of the breech block and can be assembled from either side to conform to the position of the fire and safe unit or trigger-motor. It has guides to engage corresponding grooves in the breech block, a hollow boss to house one end of the sear spring, and a bent on the front to engage the bent on the firing pin and hold this in the cocked position.

Sear retainer (figs. 9, 10 and 11)

46. The sear retainer is accommodated in a vertical recess in the rear of the breech block and has interrupted guides to engage corresponding grooves in the breech block and also shoulders on each side to position it. A transverse hole to enable the sear spring to be assembled and a vertical hole to accommodate the stem of the sear spring retainer are drilled through the sear retainer.

Sear spring retainer (figs. 9, 10 and 11)

47. The sear spring retainer consists of a head and stem, the former having guides to engage corresponding grooves in the breech block and the latter being accommodated in the sear retainer to form a bearing for the sear spring.

Sear spring retainer keeper (figs. 10 and 11)

48. The sear spring retainer keeper consists of a plate and stem. The former is accommodated in a groove on the left of the cocking lever slot in the breech block and bears on the head of the sear spring retainer to hold this in position ; it is also cut away to clear the cocking lever which holds it in the groove. The stem passes through a hole in the breech block to hold the plate longitudinally and also to form a rear bearing for the firing pin spring.

Return spring (figs. 9 and 11)

49. The return spring is mounted on a steel rod and is held in initial compression between the head of the rod, which is secured by a pin, and a collar which is retained by a stud. The head of the rod abuts against the back plate, and is held against the side of the breech casing by a dove-tailed projection, which engages a corresponding slot in the rear of the right side plate when the spring is in position in the gun.

CHAPTER 2

MECHANISM OR ACTION OF THE GUN

General

50. The parts involved in the actual working of the gun are the recoiling portions and the feed mechanism, while control of the action is effected through the rear sear release and the fire and safe unit or trigger-motor. Included in this description is the firing mechanism which is housed in the breech block, travels with it and also works independently to fire the round when the breech block is locked in the firing position.

51. Owing to the very high rate of fire of this gun all the mechanical operations are practically instantaneous, the interval between explosions being approximately 1/20th sec.

Recoil (assume the gun to have just fired) (fig. 12)

52. At the moment of the explosion, the barrel, barrel extension and breech block recoil as a single unit for about $\frac{5}{16}$ in. This ensures that the bullet has left the muzzle before the breech block is unlocked from the barrel extension. The locking-piece now rides down its cam, positive action being assured through the engagement of the prongs of the lock frame with the locking-piece pin, the breech block is freed from the barrel extension and is enabled to continue its recoil movement independently of the barrel portions. The recoil of the barrel causes the T-shaped projection on the barrel extension to engage the plunger and compress the barrel return spring; the movement, totalling about $\frac{5}{8}$ in., is arrested when the barrel extension butts against the lock-frame. At the same time the barrel extension pushes the accelerator upwards, backwards and over centre until its two ramps are interlocked with the front face of the T-shaped projection on the barrel extension. This locks the barrel extension to the lock-frame against the pressure of the barrel return spring. As the accelerator rotates, its horns strike against the rear sear bent on the breech block thus accelerating the moving block sharply to the rear.

53. The breech block in moving to the rear compresses the return spring and causes the cocking lever, through its engagement with the bracket under the top plate, to swing forward and draw back the firing pin. The sear, under the action of its spring, engages the bent of the firing pin and holds it in the cocked position. At the same time a live round, which is gripped between the claw of the transporter and the ejector, is withdrawn from the belt and carried by the

transporter to the rear. The breech block as it moves away from the barrel extracts the empty case from the chamber. On reaching the end of the rear cam on the side plate the transporter plunger is compressed and the transporter is forced down by the ramp on the breech cover. This action brings the live round into position in the cartridge guides of the breech block, while simultaneously the ejector bearing upon the empty cartridge case ejects it through the opening in the bottom of the breech casing.

Return (fig. 12)

54. The recoil of the breech block is arrested when the block strikes the buffer in the back plate. Provided that the rear sear release unit is being operated, thus depressing the sear, the block flies forward on the rebound, its movement being assisted by the return spring. During the forward movement of the breech block the live round is fed into the chamber and when the transporter plunger meets the front cam on the side plate the transporter is raised and the ejector is lifted clear of the round.

55. When the bent on the breech block strikes the horns of the accelerator the latter rotates forwards and downwards, allowing the barrel return spring to reassert itself and return the barrel extension and barrel with breech block all together into the firing position. As the barrel extension moves away from the lock-frame the locking-piece pin is disengaged from the prongs of the lock-frame and the locking-piece, riding up the incline on the locking-piece cam, engages the recess in the breech block and locks it to the barrel extension. Whilst the last movement is taking place the claw of the transporter engages the next round in the belt, under the action of the spring in the breech cover, and the ejector embraces the round ready to support it.

Feed mechanism

56. As the breech block travels backwards and forwards the feed lever is caused to oscillate by the sliding of its stud in the cam groove in the breech block. During the rearward travel of the breech block the feed lever moves the feed slide across the belt until the feed pawl, under the action of its spring, engages behind the round which is held in the feed opening by the retaining pawl. When the breech block travels forward, the feed lever swings in the opposite direction and causes the feed slide to move back and feed in the belt through the engagement of the feed pawl with the round. Double feed is prevented by the feed pawl leg and the round is brought into position against the bullet and cartridge stops ready to be gripped by the transporter. At the same time a

second round is brought over and is engaged by the retaining pawl. As each round is withdrawn by the transporter the link falls away and is ejected from the feed opening at the opposite side of the belt.

Rear sear mechanism

57. When the firing control is operated the plunger of the rear sear release unit engages the rear sear lever, causing it to rotate and depress the rear sear until the bent is below the level of the lock frame plates, thus allowing the breech block to fly freely forward.

58. When the firing control plunger is released the plunger of the rear sear release unit is withdrawn and the compressed rear sear spring raises the rear sear. As the breech block recoils, its rear end depresses and overrides the rear sear, which again rises and engages the bent of the breech block as this rebounds off the buffer in the back plate. The engagement of the two bents carries the rear sear forward which causes the rear sear cradle to rotate and compress the sear buffer spring, so absorbing the blow.

Firing mechanism

59. At the end of the forward movement of the recoiling portions the sear meets the projecting plunger of the fire and safe unit or the trigger-motor, whichever is fitted, and is forced out of engagement with the bent of the firing pin. The firing pin spring is thus released, reasserts itself and forces the firing pin forward to strike the cap of the cartridge in the chamber and fire the round. The cycle of operations is then repeated so long as any ammunition is left in the belt and the firing control is operated.

Safety arrangements

60. The following mechanical safety devices are provided to ensure that the gun cannot be fired before the breech block is locked and the recoiling portions right home.

61. The cam in the breech casing is arranged to raise the locking-piece and lock the breech block to the barrel extension before the barrel and the barrel extension reach their fully forward position.

62. The sear cannot be engaged by the plunger of the fire and safe unit or trigger-motor until the barrel and barrel extension are right home, therefore the firing pin cannot be released prematurely.

63. If the sear is accidentally released during the forward travel of the breech block, the cocking lever will take the weight of the firing pin spring and render the force of the blow insufficient to fire the cap.

CHAPTER 3

MISCELLANEOUS

Loading mechanism (fig. 3)

64. The loading mechanism has been introduced to facilitate the initial loading of the gun and the clearing of stoppages and consists of the following parts.

Loading lever with spring, bolt, washer, nut and split pin; loading lever pin; operating handle; register collars and spring.

65. The loading mechanism can be mounted on either side of the breech casing to suit the direction of feed, being held in position by the loading lever bolt which passes through the lugs on the bottom plate. The loading lever is keyed to the bolt by hexagon flats and to the operating handle by the loading lever pin. A number of holes are drilled in both the lever and the handle to enable the position of the latter to be suitably adjusted. Between the lugs of the bottom plate is positioned the register spring and washers. The washers are keyed to the loading lever bolt and, projections on their outer faces engaging corresponding recesses in the lugs, hold the loading lever in the forward position. When the operating handle is pulled to the rear the loading lever, engaging the cocking stud, withdraws the recoiling portions of the gun simulating the recoil action. The loading mechanism is returned to the forward position by the loading lever spring.

Note.—In some cases a toggle may be employed in place of the loading mechanism.

Ammunition belt and links (fig. 12)

66. The links that couple the cartridges together to form the ammunition belt are made of sheet spring steel. Each link consists of three rings of unequal size in the same plane. Two rings are on the one side and one ring on the other side of the link. The former rings are designed to fit tightly the neck and body respectively of the cartridge case, the latter ring is of larger diameter and fits loosely around the cartridge when this is in position.

67. When the belt is assembled, the single ring is positioned between the two rings of the next link, and the cartridge case, passing through the three rings, forms the axis pin of the two links, the whole constituting a hinge. Exact location is assured by a flange on the smallest ring pressing against the shoulder of the cartridge neck. When a cartridge is correctly assembled and positioned in a link the distance from the bullet edge of the smallest ring to the base of the cartridge case must

be between 2·072 in. and 2·052 in., the distance being assured by the setting of the filling machine. The force required to pull a cartridge out of a link is between 8 lb. and 16 lb. when correctly assembled.

68. The complete belt is made up of loosely linked rounds, the flexibility of the belt being aided by tapering the sides of the largest ring. When fed to the gun the belt becomes disintegrated as rounds are withdrawn and the links are thrown out of the gun to fall overboard or to be collected in a container according to the installation and type of aeroplane.

69. The links are finished with a protective coating and, when issued, are free from oil and are ready for use. Links that have been previously used must be inspected, before they are again assembled into belts, to ensure that they are not damaged or stretched. All defective links must be rei unserviceable to prevent them from being assemblec belts.

Care and cleaning

70. General instructions for the care, unit storage cleaning of machine guns are given in A.P.1641/2 and the care of barrels in A.P.1641/3.

71. Operations to be done before, between and after flights, at the 120 hour inspection of the aeroplane, and operations after 4,000 ± 500 rounds have been fired, are laid down in A.P.1641C, Vol. II, Part 3, Chap. 2.

72. The muzzle-attachment has been specially fitted in the factory and no attempt should be made to detach it in the Service.

73. It is essential that the muzzle-attachment be held with a second spanner when the flash eliminator is being removed.

74. When the gun is unloaded the return spring and firing pin spring must always be eased.

75. Do not put any strain on the barrel casing.

76. When assembling the barrel into the barrel casing take care to avoid damaging the muzzle and the barrel bearing in the trunnion block.

77. Do not clamp the gun in a vice for the purpose of cleaning, inspection or repair. An attachment for securing machine guns to the bench has been introduced and must always be used.

Belt positioning machine (fig. 13)

78. The belt positioning machine consists of a base plate which is secured to the bench by three wood screws. At one end of the plate are two vertical projections in which are cut slots to take the trunnion pins of a spur segment, at the other end are formed positioning stops for the links and in the top of the plate are a number of grooves to accommodate the ammunition.

79. A rack with cross piece slides on the top of the base plate being operated by a lever with spur segment. The movement of the rack and crosspiece is limited by a stop on its underside coming into contact with the end plate. The cross piece is prevented from lifting by guide lugs at its extremities engaging under the sides of the plate, which are cut away to enable the rack and cross-piece to be assembled.

80. The spur segment and lever are in one casting which pivots about its trunnion pins in the slots in the projections on the base plate.

81. To position the rounds, the belt is laid upon the base plate with the bullets between the positioning stops as indicated in fig. 13. The cross-piece is then brought against the base of the cartridges and the lever operated until its movement is arrested by the shoulders on the cartridges entering the bell mouth of the links. The lever is then released, the belt is moved across the base plate until the next rounds are in position and the process repeated. When moving the belt, care must be taken to ensure that no rounds are missed otherwise a stoppage may result.

82. The machine must be kept free from oil and is to be cleaned with a dry rag before use.

CHAPTER 4

LOADING, UNLOADING, STOPPAGES AND IMMEDIATE ACTION

LOADING AND UNLOADING

To make up the belt by hand

83. Arrange the links so that the single ring of one link is positioned between the two rings of the next link and with the smaller of the two rings on the same side.

84. Take a round of ammunition and push it into the three rings until the neck of the cartridge enters the smallest ring, thus coupling the links together.

85. Repeat the above operation until the required length of belt is made up.

86. Pass the belt through the belt positioning machine to ensure correct positioning of the rounds in the links.

Note.—All oil must be removed from the links by wiping them with a rag which has been dipped in grade III petrol and allowing them to dry and they must be examined for damage before being assembled into belts.

Precautions to be observed before loading the gun

87. Ensure that the cartridge and bullet stops are free and that the feed and retaining pawl springs react smartly.

88. Cock the gun by pulling the loading handle sharply to the rear and replace the loading handle in the forward position.

89. Operate the rear sear release unit by hand and if the breech block does not travel right home look for the following:—

(i) Friction due to burrs on the guides of the breech block.

(ii) Incorrect adjustment of, or burrs on, the locking-piece cam.

(iii) Bent return spring rod or weak return spring.

(iv) Weak barrel return spring.

(v) Fouling in the muzzle attachment.

(vi) Dirt or brass chips in the trunnion block.

To load the gun

90. Ensure that the firing control mechanism is set to " SAFE ".

91. When the feed is from the right the single ring of the first link must be towards the gun (*see* fig. 12A). When the feed is from the left the two rings of the first link must be towards the gun.

92. Ensure that the cocking stud is in the forward position, disengage the catch and raise the breech cover.

93. Raise the transporter, place the first round of the belt in the feed-way against the cartridge and bullet stops and ensure that the second round is engaged by the retaining pawl. It may be necessary to pull back slightly the loading handle to enable this to be done.

94. Engage the transporter with the first round and close the breech cover.

95. Cock the gun by pulling the loading handle sharply to the rear, replace the loading handle in the forward position and set the firing control mechanism to " FIRE ".

96. The gun is now ready to fire.

97. If, after cocking the gun as in para. 95 there is no round on the face of the breech block, look for the following :—

Defective transporter, or ejector.

To unload the gun

98. Ensure that the firing control mechanism is set to " SAFE ".

99. Disengage the catch and raise the breech cover.

100. Remove the belt from the feed-way.

101. Pull the loading handle right back and, whilst holding it in this position, vary the pressure on it so that the transporter can be raised by the forefinger, then remove the round from the guides on the breech block, lower the transporter and replace the loading handle in the forward position.

102. Operate the rear sear release unit by hand and then close the breech cover.

103. Cock the gun, operate the rear sear release unit by hand and ease the firing pin spring.

104. The gun is now unloaded.

STOPPAGES AND IMMEDIATE ACTION

Stoppages

105. The chief stoppages are mainly due to such defects as faulty breeching up, misalignment of chutes, faulty ammunition, short or broken firing pin, weak or broken firing pin spring, failure to feed, broken or faulty transporter, damaged cartridge guides on breech block and excessive friction. Stoppages can be reduced to a minimum by the careful observance of maintenance instructions.

106. The position of the cocking stud will indicate the action that must be taken to clear a particular stoppage.

The positions are as follows :—

First position.—At the forward end of its slot.

Second position.—From nearly right forward to half-way back in its slot.

Third position.—More than half-way back in its slot.

Immediate action

107. Immediate action is the immediate application of a probable remedy for a stoppage based on the position of the cocking stud. It must not be considered to be complete until the gun is again functioning satisfactorily. The immediate action table is set out to give a clear indication of the nature and cause of each stoppage and the probable remedy.

108. The type and number of stoppages that can be cleared in the air will depend upon the installation and the spare parts carried in the aeroplane.

Immediate action table

109.

Position of Cocking Stud.	Immediate Action.	Result.	Cause.	Remedy.
First	Cock the gun. Fire and watch feed.	(1) Gun fires.	(1) Misfire due to defective ammunition.	(1) Nil.
		(2) Feed takes place but gun does not fire.	(2) (i) Defective firing pin or spring. (ii) Dirt, congealed oil or brass chips in firing pin way in breech block. (iii) Defective firing control mechanism or insufficient pressure in the system.	(2) Unload. (i) Change breech block. Reload and continue firing. (ii) Remove and clean breech block or change breech block. Reload and continue firing. (iii) Examine and test firing control mechanism.
		(3) Feed does not take place, gun does not fire.	(3) (i) Defective transporter. (ii) Defective feed pawl spring. (iii) Defective retaining pawl spring.	(3) Unload. (i), (ii) and (iii) Change the defective part. Reload and continue firing.

Immediate action table—*continued*

Position of Cocking Stud.	Immediate Action.	Result.	Cause.	Remedy.
Second ..	Cock the gun. Fire.	(1) Gun repeats stoppage.	(1) (i) Transporter fails to lift due to a defective plunger or spring. (ii) Loose link jammed in feed-way. (iii) Cross feed due to cartridge jamming in feed-way. (*Note.*—(1) (iii) May also cause a stoppage in the first position. (iv) Separated case.	(1) Unload. (i) Change transporter. Reload and continue firing. (ii) Clear link. Ensure that feed pawl leg is not damaged. Reload and continue firing. (iii) Adjust belt. Reload and continue firing. This stoppage can be practically eliminated by correct filling and positioning of cartridges in belt. (iv) If inspection indicates that stoppage is not due to above use clearing plug to remove the separated case unless it is brought out by the round on the breech face. Reload and continue firing.

Immediate action table—*continued*

Position of Cocking Stud.	Immediate Action.	Result.	Cause.	Remedy.
Third ..	Cock the gun. Raise breech cover. Clear the gun. Operate rear sear release unit. Close the breech cover. Cock the gun. Fire.	(1) Gun fires.	(1) Defective ammunition. Empty case not extracted from chamber or ejected from breech block.	(1) Nil.
		(2) Gun repeats stoppage.	(2) (i) Defective breech block. (ii) Defective ejector.	(2) Unload. (i) Change breech block. (ii) Change transporter. Reload and continue firing.

Notes.—(i) Excessive fouling in the muzzle attachment, a defective barrel return spring or a displaced barrel locking spring will cause stoppage in the second position.

(ii) Faulty breeching up of the gun will cause a stoppage due to a separated case or to the recoiling portions not going fully forward.

(iii) A broken transporter guide spring or a bent or damaged return spring rod may cause stoppages in various positions.

Key to Illustrations

Part No.	*Name of Part.*	*Fig. No.*
1	Barrel casing	1
2	Blast tube adaptor	1, 4
3	Muzzle-attachment	1
4	Flash eliminator and split pin	1
5	Breech casing	1, 4
6	Breech cover	1, 4, 5
7	Trunnion block	1, 4, 12
8	Front mounting bracket	1
9	Loading mechanism and rear mounting bracket lugs	1
10	Back plate	1
11	Locking-piece cam, nut and split pin	1, 4, 12
12	Back plate body	2
13	Back plate catch	1, 2
14	Back plate plunger	2
15	Back plate catch spring	2
16	Back plate catch pin	2
17	Buffer tube	2
18	Buffer	2
19	Buffer discs (22)	2
20	Buffer tube screw	2
21	Buffer tube screw plunger and spring	2
22	Loading lever	3, 12
23	Loading lever bolt and nut	3
24	Loading lever pin	3
25	Loading lever washer	3
26	Operating handle	3, 4, 12
27	Register collars and spring	3
28	Loading lever spring	3, 12
29	Retaining pawl brackets, left and right	1, 4
30	Breech cover detent, spring and split pin	4
31	Breech cover axis pin and split pin	4
32	Ammunition box neck lugs (4)	1, 4
33	Filling piece	4
34	Retaining pawl	1, 4
35	Retaining pawl spring	4
36	Retaining pawl pins and split pins	4
37	Rear cartridge stop	4, 12
38	Front cartridge and bullet nose stop	4, 12
39	Front transporter cam	4
40	Rear transporter cam	4
41	Breech cover catch	4, 5
42	Breech cover catch axis pin with lever	4, 5
43	Breech cover catch spring	4, 5
44	Transporter guide spring	4, 5
45	Transporter ramp	5
46	Feed lever	4, 5
47	Feed lever plunger with spring	5
48	Feed lever bush and split pin	4, 5
49	Feed lever stud	5
50	Feed slide	4, 5
51	Feed pawl	4, 5
52	Feed pawl pin with retainer	5
53	Feed pawl leg	4, 5
54	Feed pawl leg pin	5
55	Feed pawl spring	5
56	Lock frame plates	7
57	Separator	6, 7
58	Rear sear stop	6, 7
59	Distance piece	6, 7
60	Accelerator	6, 7, 12
61	Accelerator pin with retainer	7
62	Accelerator stop	7
63	Barrel return spring plunger	6, 7
64	Barrel return spring	6, 7
65	Barrel return spring socket	7
66	Barrel return spring socket pins	7
67	Rear sear buffer spring	6, 7
68	Rear sear buffer spring washer	7
69	Rear sear cradle	6, 7, 12
70	Rear sear	6, 7, 12
71	Rear sear pin	7
72	Rear sear plunger with spring	7
73	Rear sear lever	7, 12
74	Rear sear lever pin	7
75	Lock frame retainer	6, 7
76	Lock frame retainer tube	7
77	Lock frame retainer tube stud	7
78	Lock frame retainer spring	7
79	Barrel extension	8, 12
80	Barrel locking spring	8, 12
81	Locking-piece	8
82	Locking-piece pin with retainer	8, 12
83	Barrel extension stud	8
84	Return spring with rod	9, 12
85	Cocking stud	9, 10, 11
86	Cocking stud pin	9
87	Transporter	9, 10, 11, 12
88	Transporter plunger	9, 10, 11
89	Transporter claw	9, 10, 11
90	Transporter stop	9, 10
91	Ejector	9, 11
92	Ejector pin	11
93	Ejector spring	11
94	Switch plate	9, 10, 11
95	Switch plate plunger with spring	11
96	Cocking lever	9, 10, 11, 12
97	Cocking lever pin	9, 11
98	Firing pin with spring	9, 11
99	Sear	9, 10, 11
100	Sear spring	10, 11
101	Sear retainer	9, 10, 11
102	Sear spring retainer	9, 10, 11
103	Sear spring retainer keeper	10, 11
104	Breech block	12
105	Breech end of barrel	12
106	Lock frame	12
107	Drill cartridges	12
108	Ammunition belt links	12

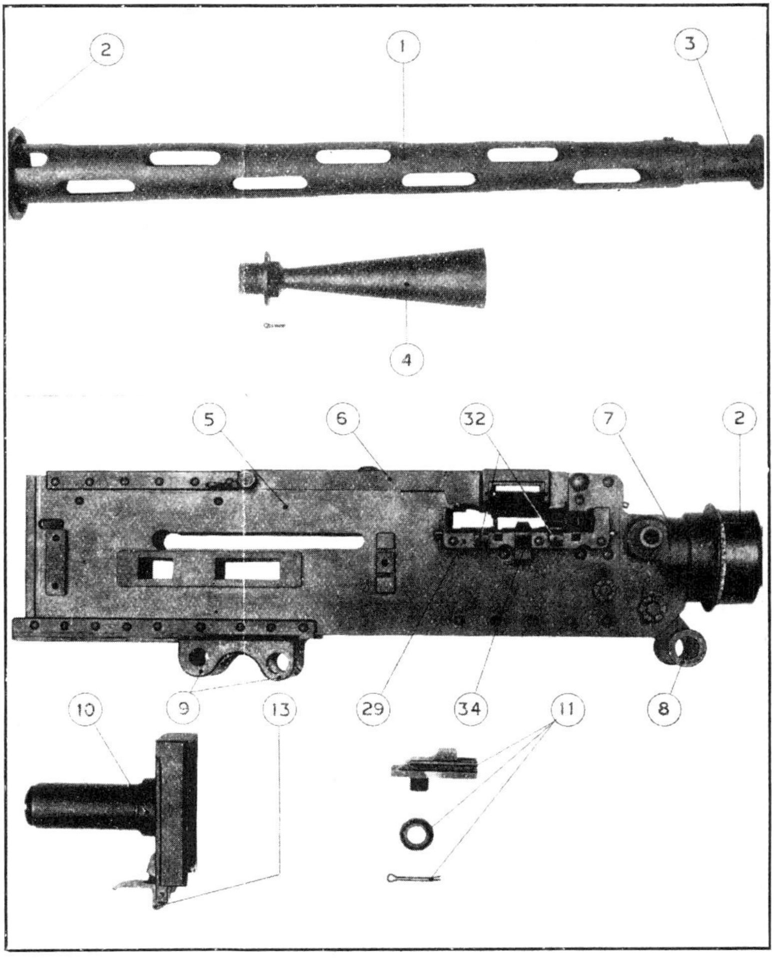

Fig. 1.—Barrel and breech casings with flash eliminator, back plate and locking-piece cam.

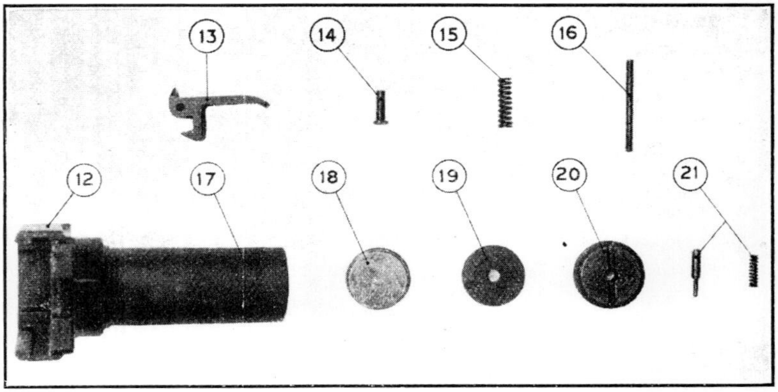

Fig. 2.—Back plate components.

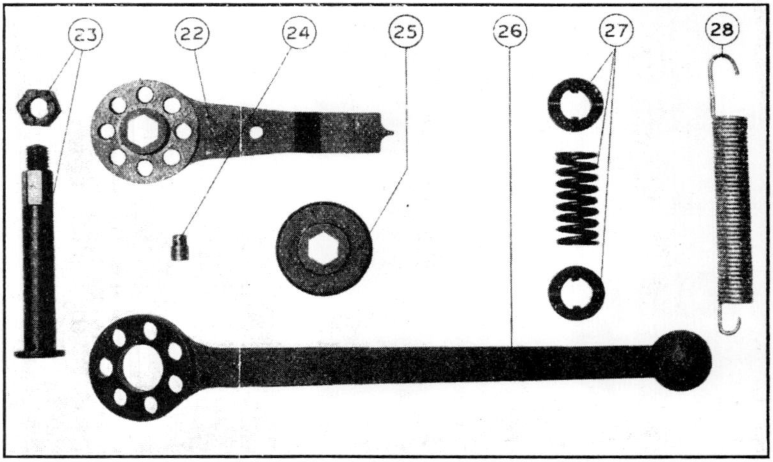

Fig. 3.—Loading mechanism.

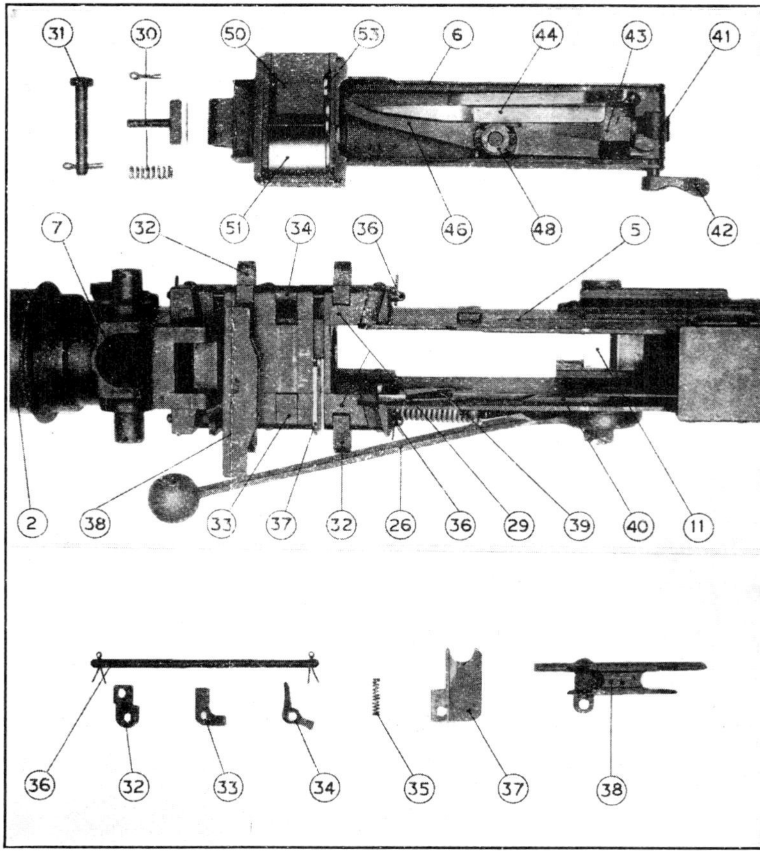

Fig. 4.—Breech cover, breech casing and feed opening components.

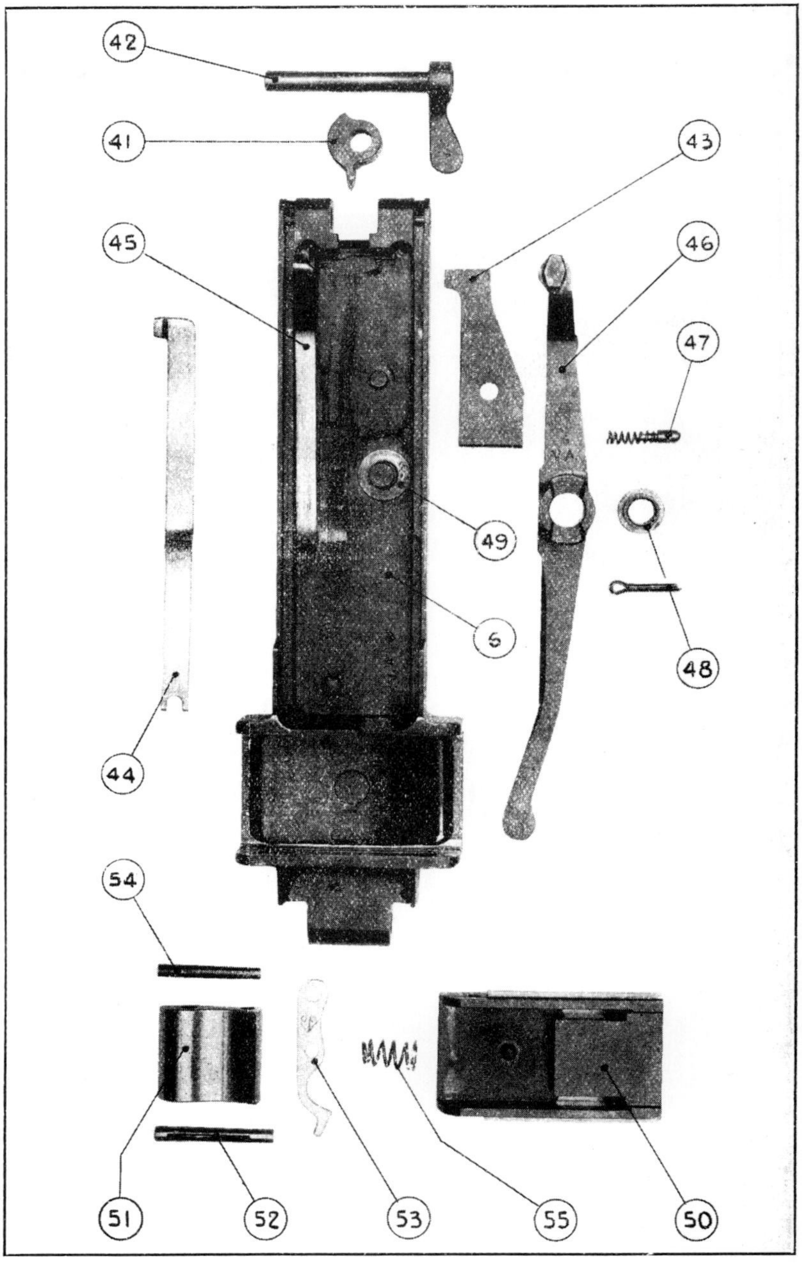

Fig. 5.—Breech cover components.

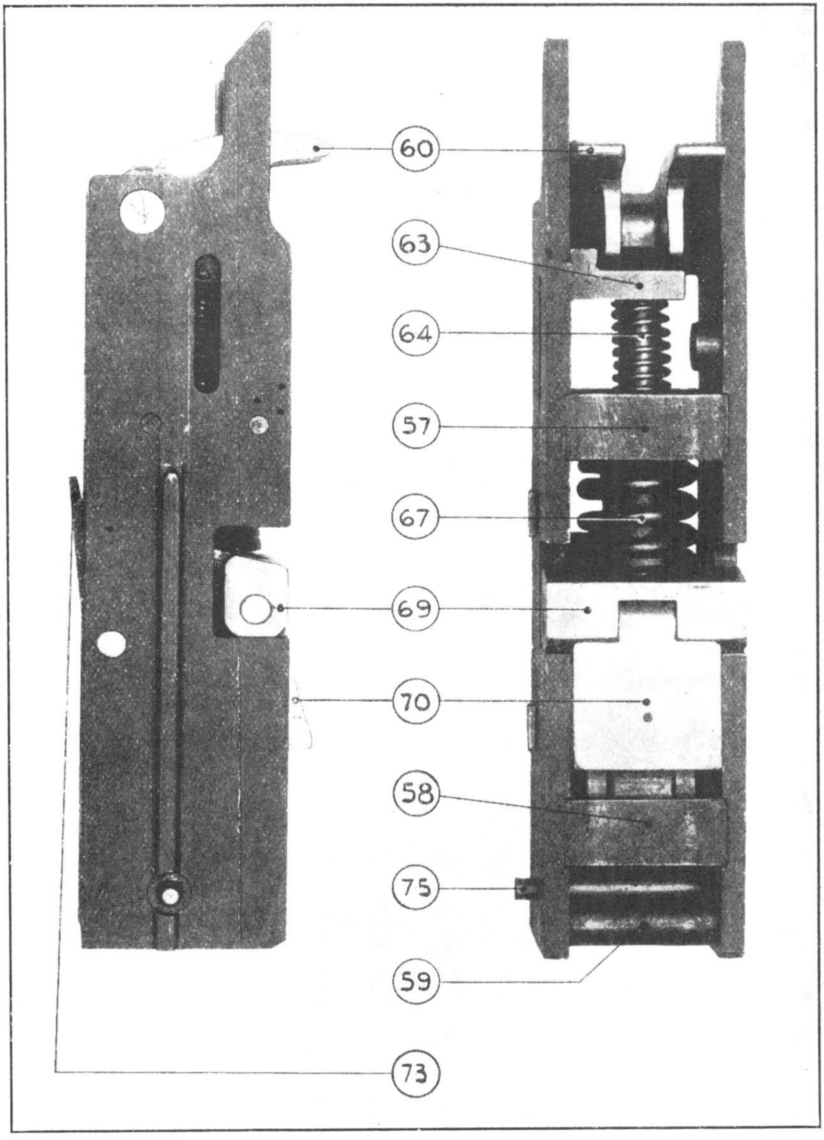

Fig. 6.—Lock frame.

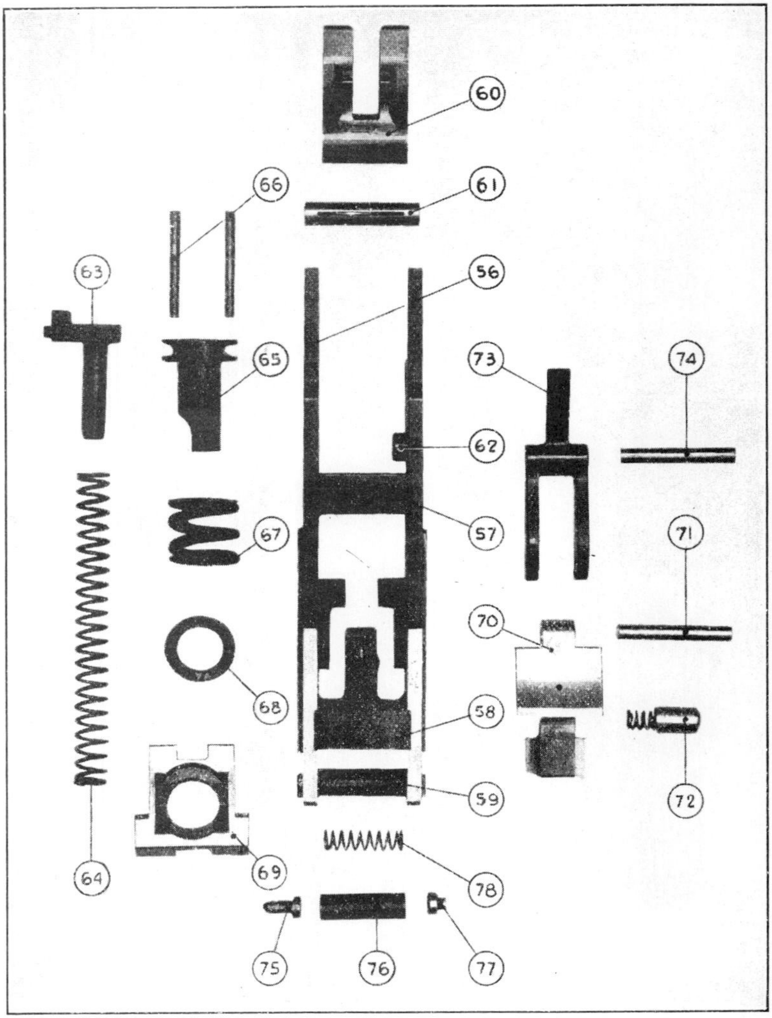

Fig. 7.—Lock frame components.

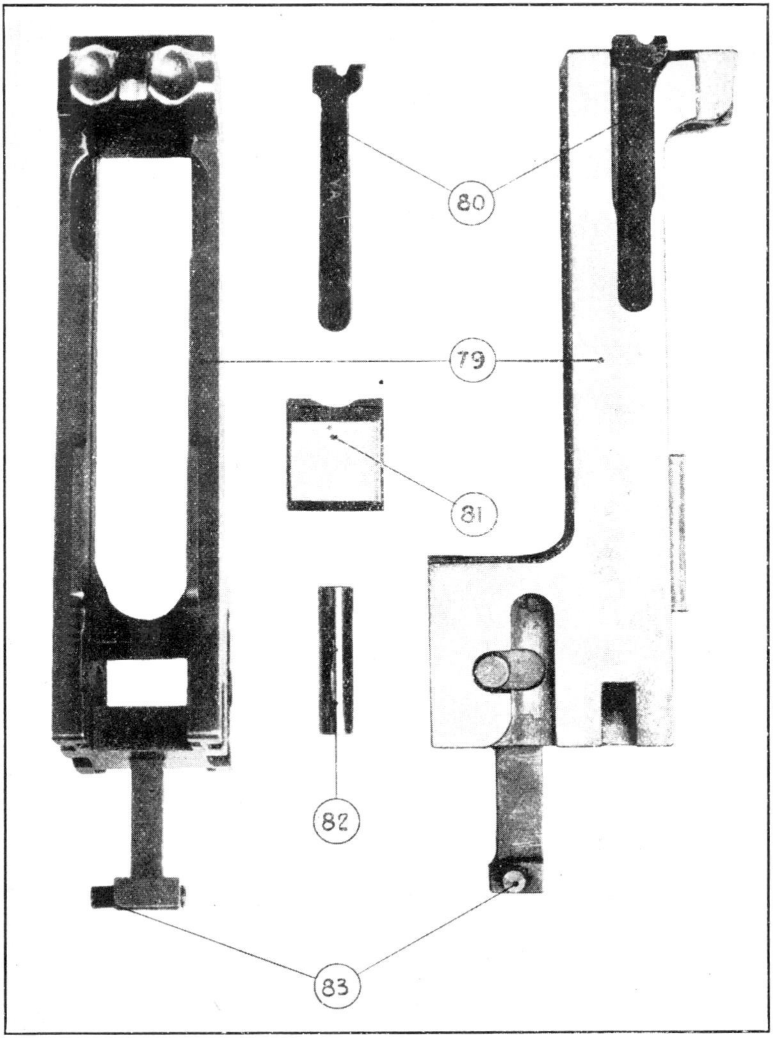

Fig. 8.—Barrel extension, side and top views and components.

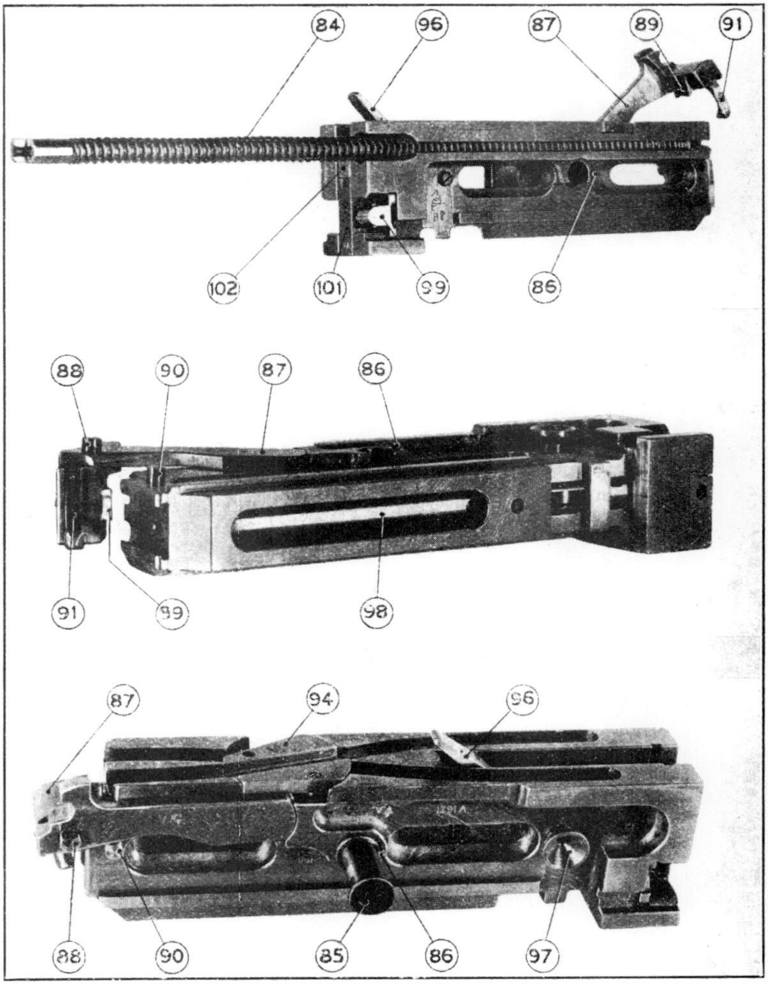

Fig. 9.—Breech block, right, bottom and left views.

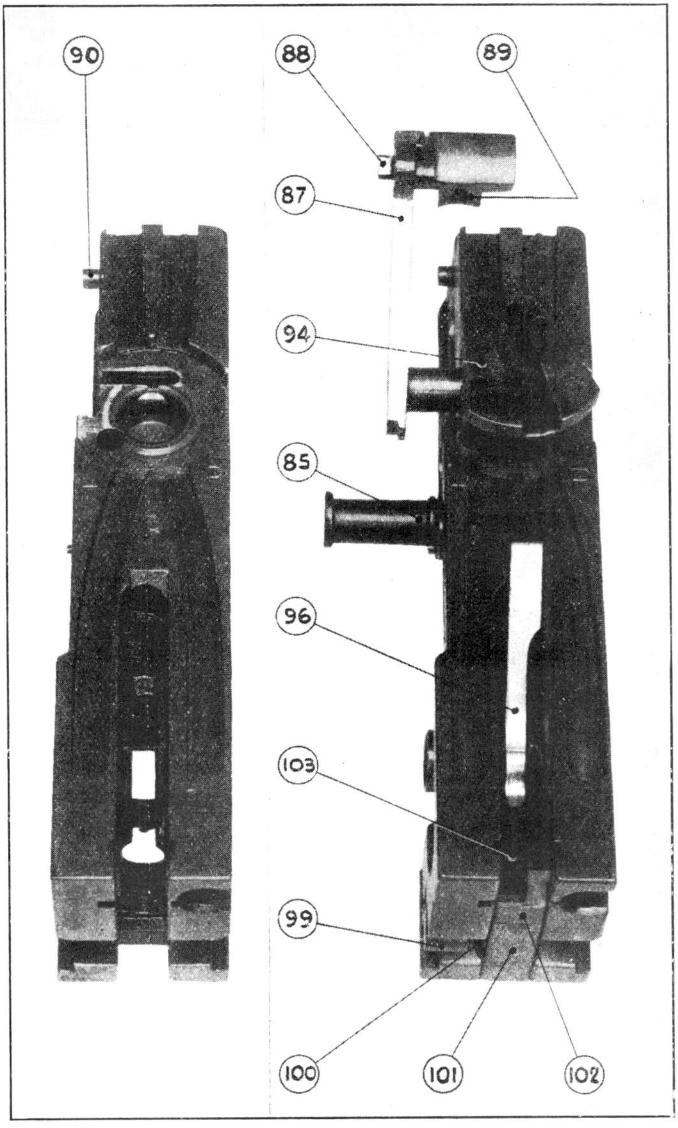

Fig. 10.—Breech block body and breech block with switch plate and transporter partly removed.

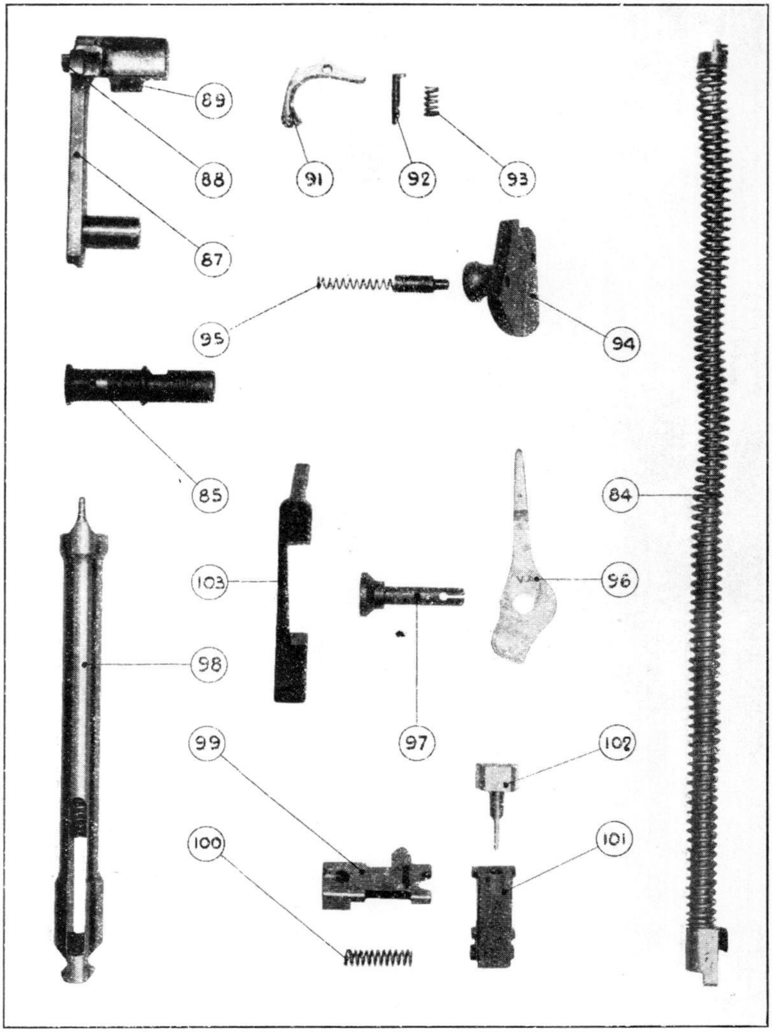

Fig. 11.—Breech block components.

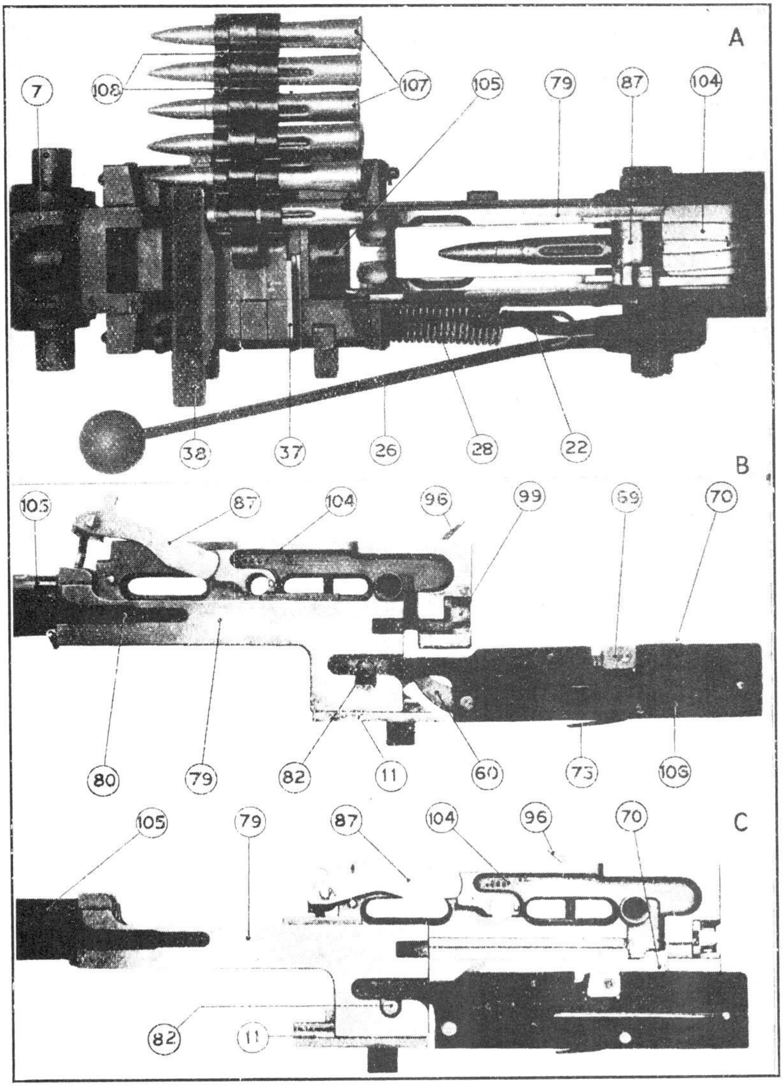

Fig. 12.—A. Breech casing with breech cover removed to show feed.
B. Recoiling portions forward with locking-piece raised.
C. Recoiling portions back with locking-piece lowered.

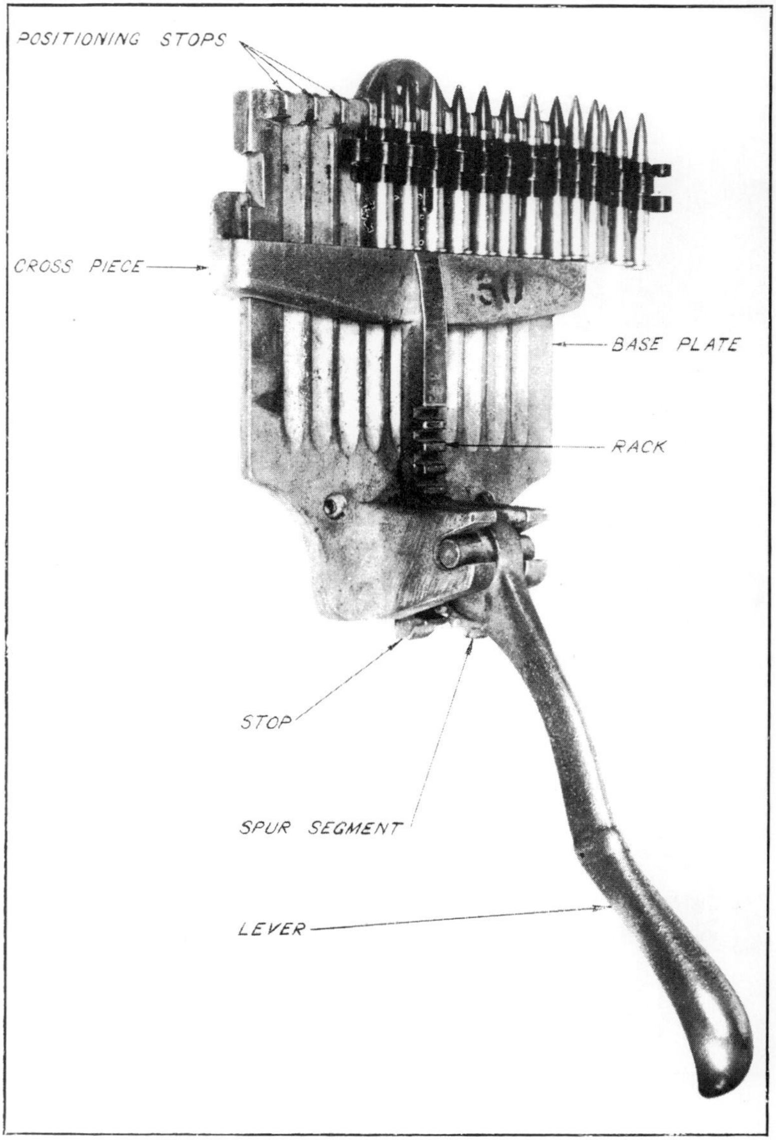

Fig. 13.—Belt positioning machine

5

Hispano 20mm Mk. I and II Guns

CONTENTS

Page

Leading particulars 7

CHAPTER 1
INTRODUCTION

Para.

General description 1

CHAPTER 2
Description of the HISPANO 20 mm. Mk. I and II GUNS

General remarks 8

Barrel with gas block and gas plug
- Barrel, Mk. I 9
- Barrel, Mk. II 12
- Gas block and gas plug 13

Recoil reducer
- Recoil reducer, Mk. I 14
- Recoil reducer, Mk. II 17

Front mounting unit
- Front mounting unit, No. 1, Mk. I 18
- Front mounting unit, No. 2, Mk. I 23

Body with locking shoulder, sear assembly, magazine carrier, catch, and ejector
- Body, Mk. I 26
- Body, Mk. II 28
- Locking shoulder, Mk. II 30
- Locking shoulder, Mk. III 31
- Sear assembly, Mk. I 32
- Sear assembly, Mk. II 36
- Magazine carrier, Mk. II 37
- Magazine catch, Mk. I 39
- Ejector, Mk. II 41
- Magazine carrier, Mk. III 43
- Magazine catch, Mk. II 44
- Ejector, Mk. III 45

Back block with return spring
- Back block, Mk. I 46
- Back block, Mk. II 48
- Return spring, Mk. II with guide and rod, Mk. II 49

Gas cylinder and piston rod assembly
- Gas cylinder, Mk. I 51
- Gas cylinder, Mk. II 52
- Piston, Mk. I 53
- Unlocking plate rods 54
- Piston, Mk. II 55

Breech block assembly 56
- Breech block, Mk. I 57
- Unlocking plates, Mk. I 59
- Locking piece 60
- Firing pin, Mk. I 62
- Extractor, Mk. II 63
- Breech block, Mk. II 65
- Unlocking plates, Mk. II 66
- Firing pin, Mk. II 67
- Extractor, Mk. III 68

Cocking units
- Pneumatic cocking unit, Mk. I 69
- Pneumatic cocking unit, Mk. II 71
- Hand cocking unit 73

CHAPTER 3

Action of the gun

	Para.
Initial cocking	74
Forward movement	77
Backward movement	80
Action of the sear	83
Action of the front mounting unit and recoil reducer	84
Action of the magazine carrier	86
Safety arrangements	87

CHAPTER 4

Description of the HISPANO MAGAZINES

Magazine No. 1, Mk. IIB	88
Front and rear plates with tie rods	89
Outer casing	90
Mouth with throat plate	91
Spring and spring casing with front plate disk and tensioning tube	92
Feed arm axis tube with feed arm bracket, feed arm plate, link, axis bush, and platform tube	93
Magazine, No. 1, Mk. IB	94
Magazine, No. 1, Mk. IA	95
Action of the magazine	96

CHAPTER 5

Installation, loading and unloading, operation, and stoppages

Installation notes	97
To load the magazine by hand	102
To unload the magazine by hand	109
To load the gun	113
To unload the gun	117
Operation	120
Stoppages	122

CHAPTER 6

Cleaning rod and tools

Cleaning rod	125
Tools	126

Fig. LIST OF ILLUSTRATIONS

Chap. 2

1. Hispano 20 mm. Mk. I gun
2. Recoil reducer and front mounting unit No. 1, Mk. I
3. Bottom view of body with locking shoulder and sear assembly
4. Top view of body with magazine carrier Mk. II, catch Mk. I, and ejector Mk. II
5. Magazine carrier Mk. III, catch Mk. II, and ejector Mk. III
6. Back block Mk. I and II, and return spring Mk. I and II
7. Gas cylinder and piston rod assembly
8. Breech block Mk. I, right side, bottom and top views
9. Breech block, Mk. I components
10. Breech block Mk. II, left side and bottom views
11. Breech block, Mk. II components
12. Pneumatic cocking unit Mk. I and hand cocking unit
13. Pneumatic cocking unit Mk. II, sectional diagram

Chap. 4

14. Magazine, No. 1, Mk. IIB components
15. Magazine in position on the gun

Chap. 6

16. Cleaning rod
17. Tools

LEADING PARTICULARS

Calibre	0·787 in.
Length of gun (overall)	8 ft. 2½ in.
Length of barrel with recoil-reducer	6 ft. 3 in.
Length of barrel	5 ft. 7¼ in.
Weight of gun with front mounting unit	109 lb.
Rifling—	
(i) Type	Concentric right-hand
(ii) Twist	1 in 53 calibres
(iii) Number of grooves ...	9
(iv) Length	5 ft. 9 in.
Ammunition—	
(i) Type	Ball and H.E.
(ii) Weight of one complete round	9 oz.
Method of feed	Magazine
Capacity of magazine	60 rounds
Weight of magazine (empty) ...	22 lb. 3 oz.
Weight of magazine (full) ...	56 lb.
Muzzle velocity	2,880 f.s.
Rate of fire (approx.)	650 rounds per min.
Marks on gun—	
(i) On the top of the body	Nature, mark, and number of the gun Makers' initials
(ii) On the barrel	Number of the barrel Proof mark
(iii) On the breech block ...	Number of the breech block
(iv) On various components	Broad arrow Viewers' marks Makers' trade mark
Sights	Separate store.

CHAPTER 1

INTRODUCTION

General description

1. The Hispano 20 mm. gun is automatic in action and, so long as the firing control mechanism is operated, should continue to fire until all the available ammunition in the magazine is expended. Its rate of fire is high, being equal to that of some of the earlier machine guns. Ammunition is fed to the gun from a spring-loaded drum-shaped magazine, which can be easily removed and exchanged when empty.

2. To enable the first round to be loaded and fired, the breech is opened and the gun cocked by a cocking unit. The breech block is held in the open position against the pressure of a return spring by the sear and, when the firing control mechanism is operated, a return spring drives the breech block forward. The breech block, as it travels forward, loads a round from the magazine, and the final action of the mechanism locks the breech and fires the gun.

3. After the first round has been loaded and fired, the gun depends for its motive power upon gas pressure. A portion of the propelling gases is deflected through a vent in the barrel on to a piston, which unlocks the breech block and allows the residual pressure in the bore to act against the empty cartridge case, and drive the breech block to the rear. The breech block, in travelling to the rear, extracts and ejects the empty cartridge case, and compresses the return spring. The rearward movement of the breech block is arrested by a buffer and, if the firing control mechanism is still operated, the return spring will drive the breech block forward to load and fire the gun again. The breech, at the moment of firing, is locked by a locking-piece which is situated at the rear of the breech block and engages a locking shoulder in the gun body.

4. The firing of the gun may be controlled by a hand-operated bowden cable, or by a pneumatically operated gun firing control mechanism, which are described in A.P.1641E, Vol. I.

5. When installed in the aeroplane the gun is cooled by air flow over the barrel.

6. The gun may be installed in any one of three positions in an aeroplane, namely in the wing, fuselage or in a turret, or can be mounted to fire from the ground. The type of mounting employed depends upon the installation and in some aeroplanes the gun may be mounted on its side. The shock of discharge is absorbed by a spring which also returns the gun to the firing position, the movement being cushioned by the action of a pneumatic piston. In some installations a recoil reducer is fitted to the muzzle of the gun.

7. The Mk. I and II guns are similar in construction and some of the components of the Mk. II gun have now been fitted to the Mk. I gun. The main parts of the Mk. I gun are the Mk. I barrel, Mk. I body, with Mk. II locking shoulder and Mk. I back block. The main parts of the Mk. II gun are the Mk. II barrel, Mk. II body with Mk. III locking shoulder and Mk. II back block.

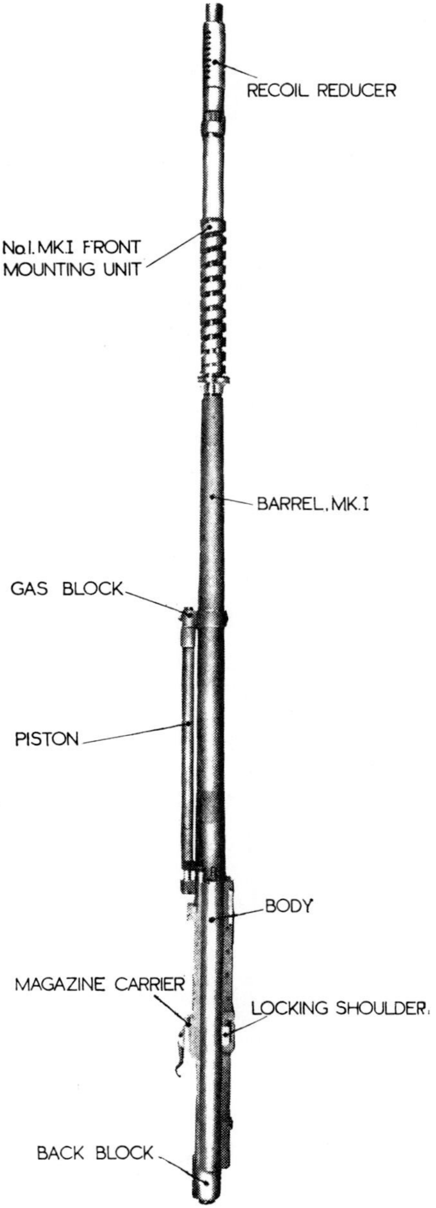

Fig. 1.—Hispano 20 mm. Mk. I gun

CHAPTER 2

Description of the HISPANO 20 mm. Mk. I and Mk. II GUNS

General remarks

8. For the purpose of description, the gun has been divided into the following groups:—

(i) Barrel, with gas block and gas plug.
(ii) Recoil reducer assembly.
(iii) Front mounting unit.
(iv) Body, with locking shoulder, sear assembly, magazine carrier, catch, and ejector.
(v) Back block with return spring and guide.
(vi) Gas cylinder and piston assembly.
(vii) Breech block assembly.
(viii) Cocking units, pneumatic and hand.

Barrel with gas block and gas plug

Barrel, Mk. I, fig. 1

9. The Mk. I barrel is bored and rifled to suit the driving band of the projectile. At the rear, the bore is increased in diameter to form a chamber which is coned to conform with the contour of the brass cartridge case. Externally the barrel is reduced in diameter from the breech towards the muzzle. It is threaded at the breech end for attachment to the body and at the muzzle for the attachment of the recoil reducer or muzzle thread protector. The barrel screws into the body and is secured by a retainer, in the underside of the body, which enters a hole in the barrel. The retainer is held by a split pin and is drilled and tapped to facilitate removal. The breech face of the barrel is recessed to clear the extractor.

10. To the rear of the muzzle thread are splines which engage the sleeve of the front mounting unit or the Mk. I retainer for the recoil reducer or muzzle thread protector, whichever is fitted. An annular groove towards the front of the splines accommodates the lock-ring of the Mk. I retainer.

11. A small gas hole or vent is drilled radially through the top of the barrel approximately 2 ft. from the breech end. In front of the gas vent the barrel tapers to a shoulder and is then parallel as far as the splines. The front mounting unit components, when fitted, are accommodated on the parallel portion.

Barrel, Mk. II

12. The Mk. II barrel differs from the Mk. I barrel in external dimensions only.

Gas block and gas plug, fig. 1 and 7

13. The gas block is shrunk and taper pinned to the barrel, a hole in it coinciding with the gas vent in the barrel. The front of the gas block is closed by the gas plug which screws into the gas cylinder, a copper washer between the block and the plug making a tight joint. The top of the gas block is closed by a screw, the removal of which enables the gas vent in the barrel to be cleaned. The gas plug and screw are secured by a piece of wire. The rear of the gas block is shaped to accommodate a gas cylinder.

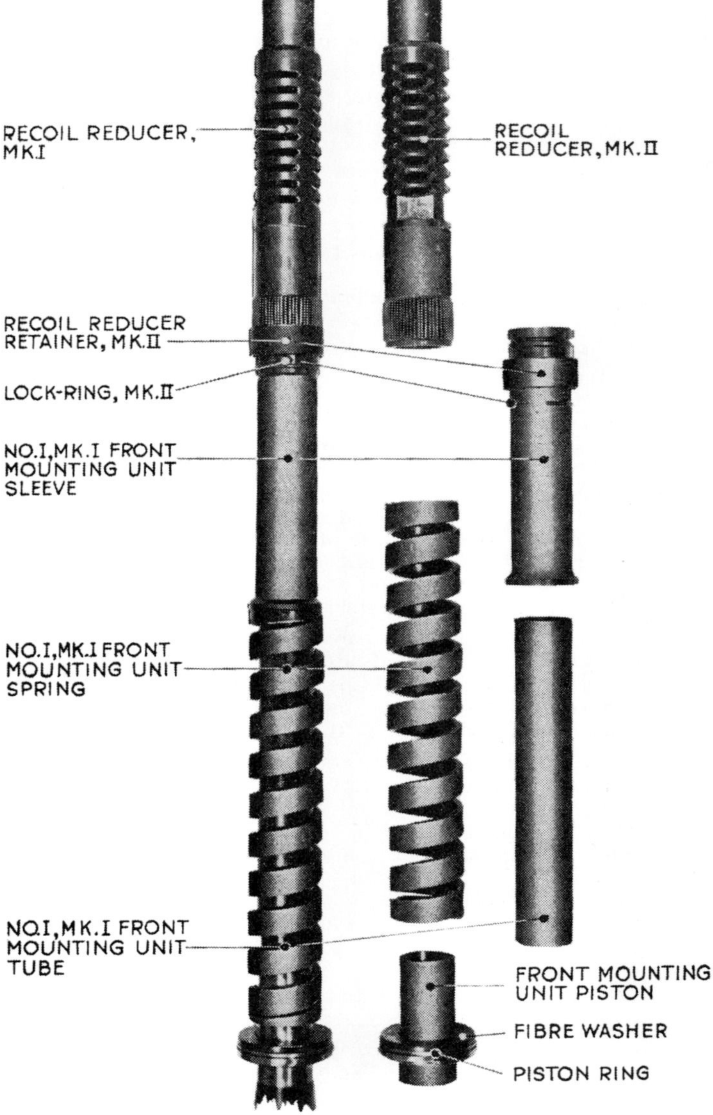

Fig. 2.—Recoil reducer and front mounting unit No. 1, Mk. I

Recoil reducer

Recoil reducer, Mk. I, fig. 2

14. The Mk. I recoil reducer consists of a cylinder with baffles, choke, adapter, and recoil reducer retainer. The rear of the cylinder screws into the adapter, the two being locked by a washer, whilst the choke screws into the front of the cylinder. The baffles are accommodated within the cylinder. A number of ports are cut in the baffles and cylinder and, being at an angle of 45 deg. to the axis of the bore, deflect the blast gases to the rear. The adapter screws on to the muzzle and is prevented from unscrewing by the recoil reducer retainer which is secured by a lock-ring. The exterior of the adapter is splined for engagement with the splines in the retainer.

15. There are two marks of recoil reducer retainer, the Mk. II for use with the No. 1, Mk. I front mounting unit, and the Mk. I for use when the front mounting unit is not fitted. The Mk. II retainer is keyed to the No. 1. Mk. I front mounting unit sleeve and is splined to the adapter, the Mk. II lock-ring engaging a groove in the sleeve behind the retainer. The Mk. I retainer is splined to both the muzzle and the adapter, the Mk. I lock-ring engaging the groove in the barrel splines.

16. The recoil reducer may be used with or without the No. 1, Mk. I front mounting unit, depending upon the installation.

Recoil reducer, Mk. II, fig. 2

17. The Mk. II recoil reducer differs from the Mk. I recoil reducer in having the baffles integral with the cylinder.

Front mounting unit

Front mounting unit No. 1, Mk. I, fig. 2

18. The No. 1, Mk. I front mounting unit consists of a piston complete with split piston ring and fibre washer, spring, tube, and sleeve which are accommodated on the muzzle of the gun between the shoulder and the recoil reducer.

19. The piston abuts the shoulder on the barrel and is accommodated within a cylinder which forms part of the front mounting in the aeroplane. Its function is to cushion the movement of the gun during run-out. The piston is held in position by the front mounting unit tube and sleeve which butt against each other, the sleeve being retained by the recoil reducer. The function of the tube is to ensure that the piston recoils with the gun and also to form a lining for the spring.

20. The No. 1, Mk. I front mounting unit spring surrounds the tube and bears between the rear of the sleeve and the front of the mounting cylinder. Initial compression is applied to the spring, through the medium of the sleeve, by the recoil reducer when assembled to the gun.

21. The No. 1, Mk. I front mounting unit sleeve is flanged at the rear to form a bearing for the spring. At the front, the interior is splined to engage the splines on the muzzle, the exterior having keyways to accommodate keys in the recoil reducer retainer and prevent this from turning. A circumferential groove accommodates the lock-ring which holds the retainer in place.

22. The No. 1, Mk. I front mounting unit is never fitted without the recoil reducer.

Front mounting unit No. 2, Mk. I

23. The No. 2, Mk. I front mounting unit is generally similar to the No. 1, Mk. I front mounting unit, but the piston and fibre washer are the only components that are interchangeable. In addition to the piston and fibre washer, the mounting unit consists of a double-rate recoil spring, distance collar, tube and sleeve.

24. The position of the piston, tube, and sleeve is the same as in the No. 1, Mk. I front mounting unit, but they are held in position by the muzzle thread protector. The double-rate recoil spring bears between the distance collar, which abuts the mounting cylinder in the aeroplane, and the rear of the sleeve. The muzzle thread protector screws on to the muzzle and is retained in a manner similar to the recoil reducer by the Mk. II muzzle thread protector retainer with Mk. II lock-ring.

25. The recoil reducer is not used with the No. 2, Mk. I front mounting unit.

Body with locking shoulder, sear assembly, magazine carrier, catch, and ejector

Body, Mk. I, fig. 3 and 4

26. The Mk. I body is generally rectangular in shape, the sides and top are solid, and integral with the right-hand side is a cocking cylinder. The bottom of the body is partially closed by the Mk. I bottom plate. At the front end the body is threaded internally for the attachment of the barrel, a vertical hole being drilled from the underside for the barrel retainer. On the top of the body is a boss which is threaded to receive the piston guide. On each side, and below the boss, longitudinal holes are drilled through the front of the body to accommodate the unlocking plate rods. To the rear of the boss is an orifice which accommodates the mouth of the magazine, there being guides, one on each side of the orifice, for the magazine carrier. Bearing ribs integral with the sides of the body are provided at the front to support the gun in the rear mounting.

27. The body is recessed internally to accommodate the breech mechanism, a featherway in the right-hand side breaking through into the rear half of the cocking cylinder to enable the cocking piston to engage the right-hand unlocking plate of the breech mechanism. The rear of the body is closed by the Mk. I back block which has square cut guides seating in guide-ways and is retained by a catch. The underside of the body is open at the front to permit the empty cartridge cases to be ejected. At the rear it is closed by the Mk. I bottom plate which accommodates the sear assembly. The breech block guides are secured by bolts to each side of the body above the ejector opening. The guides support the breech block in the firing position and are provided with ramps at the rear to engage corresponding ramps on the breech block locking-piece thus lowering it to the locked position, in front of the locking shoulder, when the breech is closed. A transverse slot is cut in each side of the body to the rear of the ejector opening, to accommodate the locking shoulder.

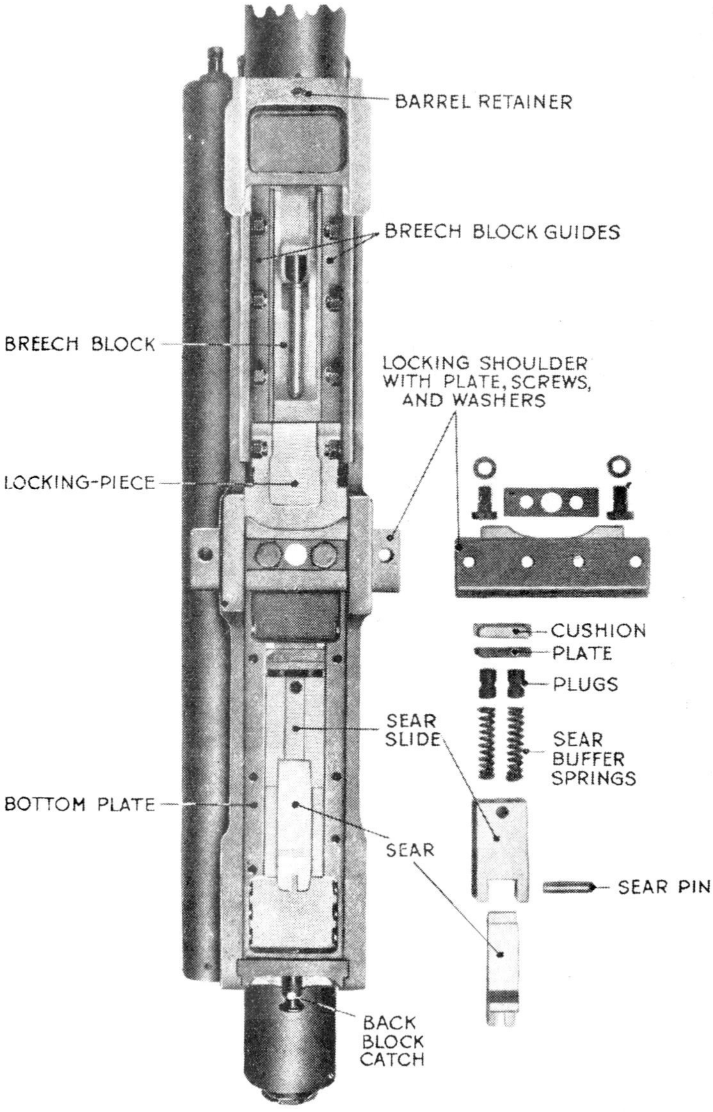

Fig. 3.—Bottom view of body with locking shoulder and sear assembly

Body, Mk. II

28. The Mk. II body differs from the Mk. I body as follows:—

(i) It is 5 mm. longer.

(ii) The guides for the Mk. II back block are buttressed instead of being square cut.

(iii) Strengthening webs are formed on the top and bottom edges.

(iv) A Mk. II bottom plate and Mk. II breech block guides are fitted.

29. The Mk. II bottom plate differs from the Mk. I bottom plate in being 5 mm. longer, whilst the Mk. I and Mk. II breech block guides differ in manufacturing tolerances only.

Locking shoulder, Mk. II, fig. 3

30. The Mk. II locking shoulder is a rectangular plate, the rear face of which is radiused. The top of the front face is bevelled slightly to allow the locking-piece to hinge down and engage in front of it. Below the bevel are two projections which support the locking-piece when it is lowered to the locked position, the underside of these projections being radiused. The locking shoulder is held in position in the slots in the body by a plate placed between the sides of the body and secured to the locking shoulder by two screws with single coil spring washers, the screws being secured by a locking wire. The ends of the locking shoulder project beyond the body and are used for attaching the gun to the rear mounting. The Mk. I locking shoulder did not project and is no longer used.

Locking shoulder, Mk. III

31. The Mk. III locking shoulder differs from the Mk. II locking shoulder in having the face of the two projections, upon which the locking-piece rests when lowered, cut at a slightly different angle.

Sear assembly, Mk. I, fig. 3

32. The Mk. I sear assembly consists of a Mk. I sear, Mk. I sear slide, sear buffer springs, plugs, plate, and cushion. The assembly is housed in the bottom plate, being retained by the pressure of the sear buffer springs. It is masked by the plate of the trigger mechanism, which also carries the sear plunger and spring, when this is assembled to the gun.

33. The Mk. I sear is hinged to the rear of the sear slide and has a bent formed on the top. The front of the bent is bevelled to allow the breech block to override and depress the sear when travelling to the rear after the firing control mechanism has been released. Below the bent, and projecting to the rear, is a fork which is engaged by the trigger when the trigger mechanism is fitted. In front of the fork is a ramp which engages a cam in the trigger mechanism plate when the breech block forces the sear forward as the bent of the locking-piece engages the bent of the sear.

34. The Mk. I sear slide is a rectangular plate having a slot at the rear to accommodate the front of the sear, which is attached by a Mk. I sear pin. At the front, two holes are drilled in the slide to house the sear buffer springs and plugs. The plugs are cylindrical in shape, and have a circumferential groove midway to accommodate the dismantling tool, which is passed through a vertical hole in the slide and holds the springs under compression when dismantling and assembling.

35. The sear buffer springs plate and cushion, the latter being of fibre, are accommodated at the front of the housing in the bottom plate, and provide a front abutment for the sear buffer springs and plugs. The steel plate must always be assembled at the rear of the fibre cushion. The function of the sear buffer springs is to absorb the shock when the bents of the breech block and sear engage.

Sear assembly, Mk. II

36. The Mk. II sear assembly is the same as the Mk. I sear assembly, except for the Mk. II sear and the Mk. II sear slide, which are differently shaped at the hinge, and the Mk. II sear pin, which has a shoulder at one end.

Magazine carrier, Mk. II, fig. 4

37. The Mk. II magazine carrier is a steel forging having an orifice to coincide with that in the body. On each side is a guide to mate with the corresponding guide-ways on the body. At the front of the carrier, two hook-shaped projections are formed to accommodate two pins at the front of the magazine mouth and hold it in position. An arm is attached to the front of the carrier by a screw with a retainer and split pin, a key on the arm mating with a key-way at the side of the carrier. The arm can be fitted to the right or left of the carrier as required. The lower end of the arm registers in the rear mounting and prevents the magazine carrier from moving when the gun recoils and runs-out. To ensure correct registering of the arm, lines are engraved on the magazine carrier and on the gun body and these lines should be made to coincide when the gun is mounted. At the rear, the carrier is shaped to accommodate the Mk. I magazine catch and Mk. II ejector.

38. The Mk. I magazine carrier differs from the Mk. II magazine carrier in that provision is made for attaching the arm to the left-hand side of the carrier only.

Magazine catch, Mk. I, fig. 4

39. The Mk. I magazine catch fits on top of the Mk. II ejector, the two being housed in guide grooves at the rear of the magazine carrier. A guide is formed on each side of the catch to mate with the guide grooves. The catch is cut away on the underside at the front to engage the catch piece at the rear of the magazine. Two ramps are formed on the top to facilitate assembling the magazine. Two holes are drilled in the rear of the catch to accommodate the magazine catch springs which bear against a vertical projection at the rear of the ejector when assembled.

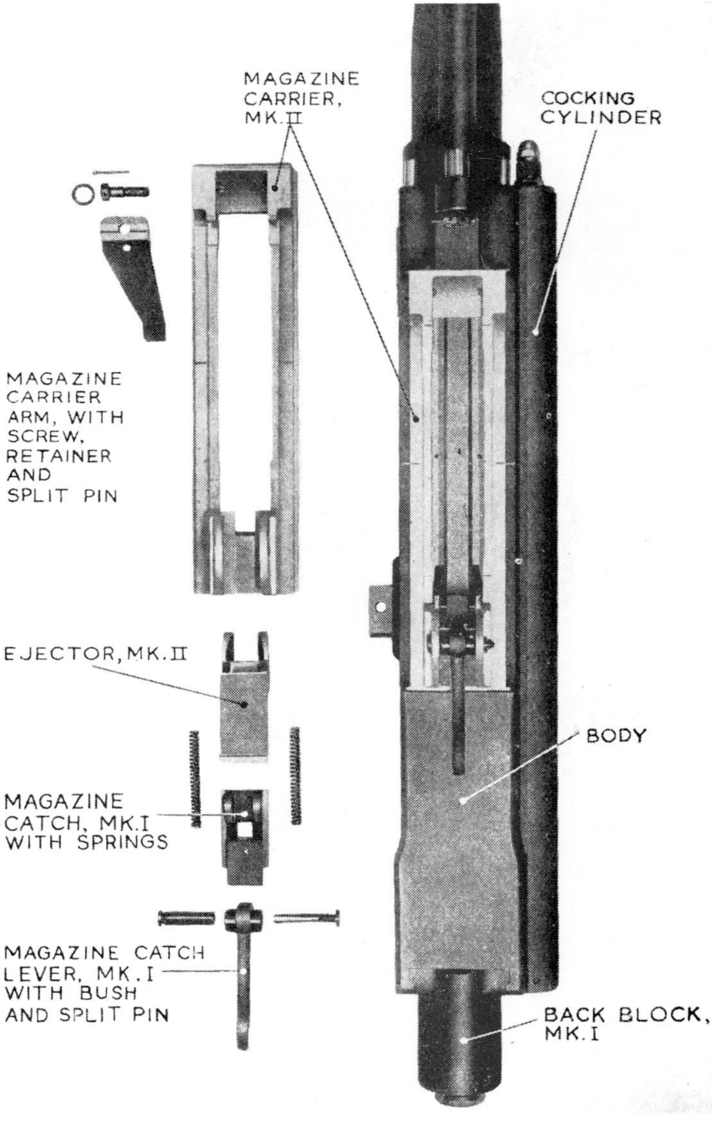

Fig. 4.—Top view of body with magazine carrier Mk. II, catch Mk. I, and ejector Mk. II

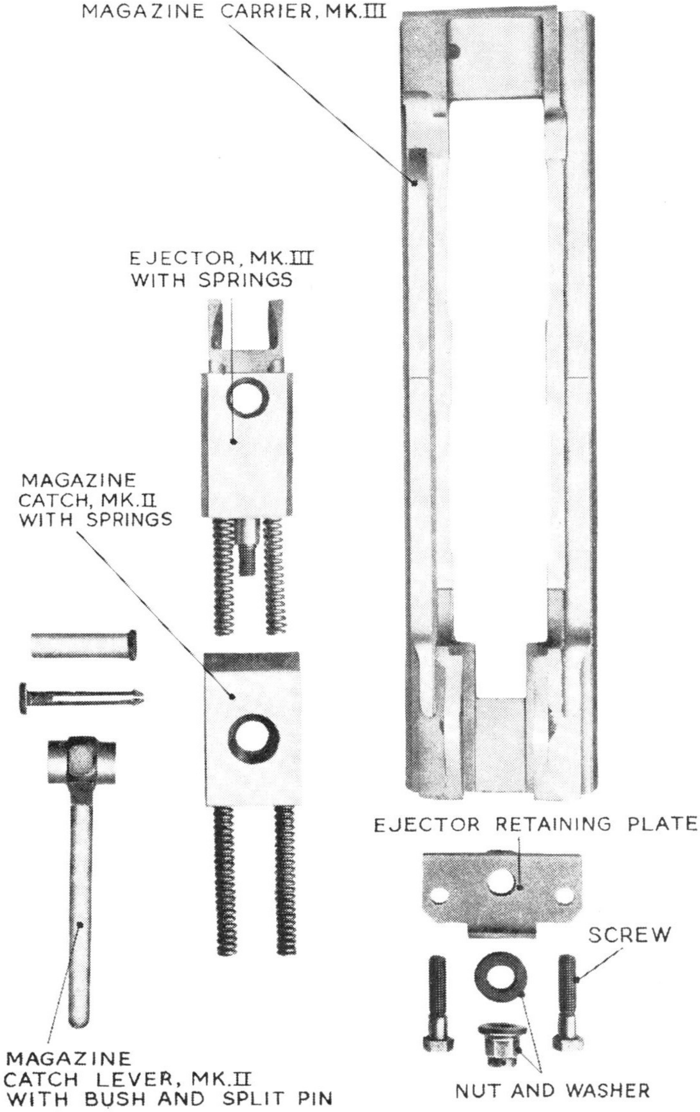

Fig. 5.—Magazine carrier Mk. III, catch Mk. II, and ejector Mk. III

40. A rectangular recess is cut near the middle of the catch to accommodate a lug on the Mk. I magazine catch lever. The lever pivots between two brackets at the rear of the magazine carrier and is held in position by a bush and split pin which pass through the brackets.

Ejector, Mk. II, fig. 4

41. The Mk. II ejector consists of two horns integral with a steel plate. The plate is held in position in the rear of the magazine carrier by the extractor, the springs of which bear against a projection at the rear of the plate. The horns project into the orifice in the body and fit one on each side of the top of the breech block and meet the base of the cartridge case when the breech block is driven to the rear.

42. The Mk. I ejector differs from the Mk. II ejector in manufacturing tolerances only.

Magazine carrier, Mk. III, fig. 5

43. The Mk. III magazine carrier is similar to the Mk. II magazine carrier but is designed to accommodate the Mk. II magazine catch and Mk. III ejector, and must be used with the Mk. II breech block.

Magazine catch, Mk. II, fig. 5

44. The Mk. II magazine catch differs from the Mk. I magazine catch as follows:—

- (i) It has no guides for engagement with the guide grooves in the magazine carrier.
- (ii) There are no ramps at the front, and the front of the catch is bevelled at the top.
- (iii) The recess for the magazine catch lever is circular, the lug of the Mk. II magazine catch lever being shaped to correspond with the hole.

Ejector, Mk. III, fig. 5

45. The Mk. III ejector is somewhat similar to the Mk. II ejector, but is buffered. The plate of the ejector is thicker and has two holes drilled in the rear for the ejector springs. Between the holes there is a threaded stud over which fits a magazine carrier back plate or ejector retaining plate. This plate is held by a nut and forms the rear abutment for both the ejector springs and the magazine catch springs. The ejector and magazine catch are housed in guide grooves in the rear of the Mk. III magazine carrier, being held in position by two screws which pass through the magazine carrier back plate and screw into the magazine carrier.

Back block with return spring

Back block, Mk. I, fig. 6

46. The Mk. I back block can only be fitted to a Mk. I body and consists of a rectangular plate, integral with which is a cylinder. On the sides of the plate are square-cut guides which mate with guide grooves in the body. The back block is retained

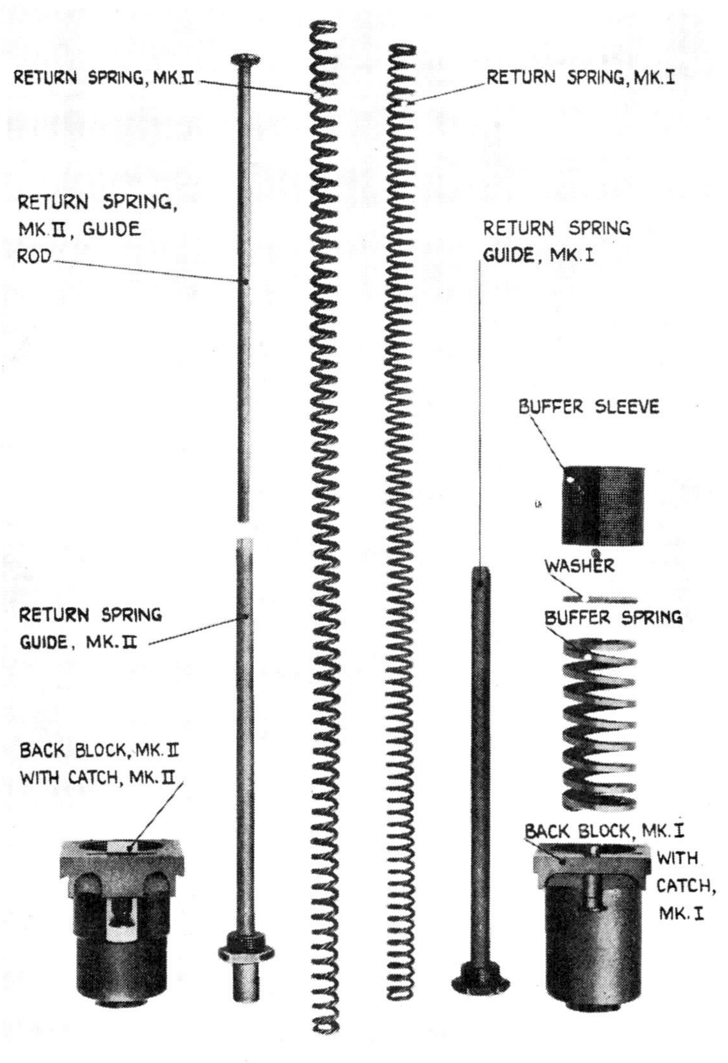

Fig. 6.—Back block Mk. I and II, and return spring Mk. I and II

in position by a catch situated at the bottom of the plate and engaging a recess in the body. The catch consists of a plunger with head and spring, the head being pinned to the plunger.

47. The cylindrical portion of the back block, which is partly closed at the rear, houses a buffer spring which is placed under initial compression by a screwed buffer sleeve, the outer end of which has an internal flange. Situated between the buffer spring and the flange is a washer, which takes the blow when the breech block is driven to the rear. The rear of the cylinder is threaded for the return spring guide.

Back block, Mk. II, fig. 6

48. The Mk. II back block differs from the Mk. I back block in having buttressed guides instead of square-cut guides, and can only be fitted to a Mk. II body. In addition it is fitted with a Mk. II catch, the end of which is rectangular.

Return spring, Mk. II, with guide and rod, Mk. II, fig. 6

49. The Mk. II return spring is a three-ply twisted coil spring and passes through the buffer spring. The Mk. II return spring guide is hollow and is provided with a nut at the rear which screws into the rear of the back block and forms an abutment for the spring. The return spring guide rod is accommodated in the breech block, its front end, which is flanged to form a front abutment for the spring, bearing against the rear of the firing pin. The rear of the rod is hollow to accommodate the split end of the assembling tool. When the breech block moves to the rear, the rod telescopes into the guide and compresses the return spring.

50. The Mk. I return spring and Mk. I guide, are obsolescent and all future Mk. I and II guns will be fitted with the Mk. II pattern return spring and guide.

Gas cylinder and piston rod assembly

Gas cylinder, Mk. I, fig. 7

51. The Mk. I gas cylinder, which accommodates the piston head, is spigoted into the gas block, and is threaded at the front to receive the gas plug which holds it in position. The cylinder is prevented from turning by a projection which is shaped to fit against the barrel.

Gas cylinder, Mk. II

52. The Mk. II gas cylinder differs from the Mk. I gas cylinder in having hexagon flats instead of the projection to prevent it from turning.

Piston, Mk. I, fig. 1 and 7

53. The piston head is recessed at the front end, has circumferential grooves cut in its exterior, and is accommodated in the gas cylinder. It is a press fit in the end of the piston rod. The piston is hollow and accommodates a spring which returns it to the front after firing. At the rear, the piston is supported on a guide which screws into a boss at the front of the body and is secured by a retainer and split pin. The guide also forms a rear abutment for the piston spring. At the rear, the piston is shaped to form a fork which bears against the unlocking plate rods.

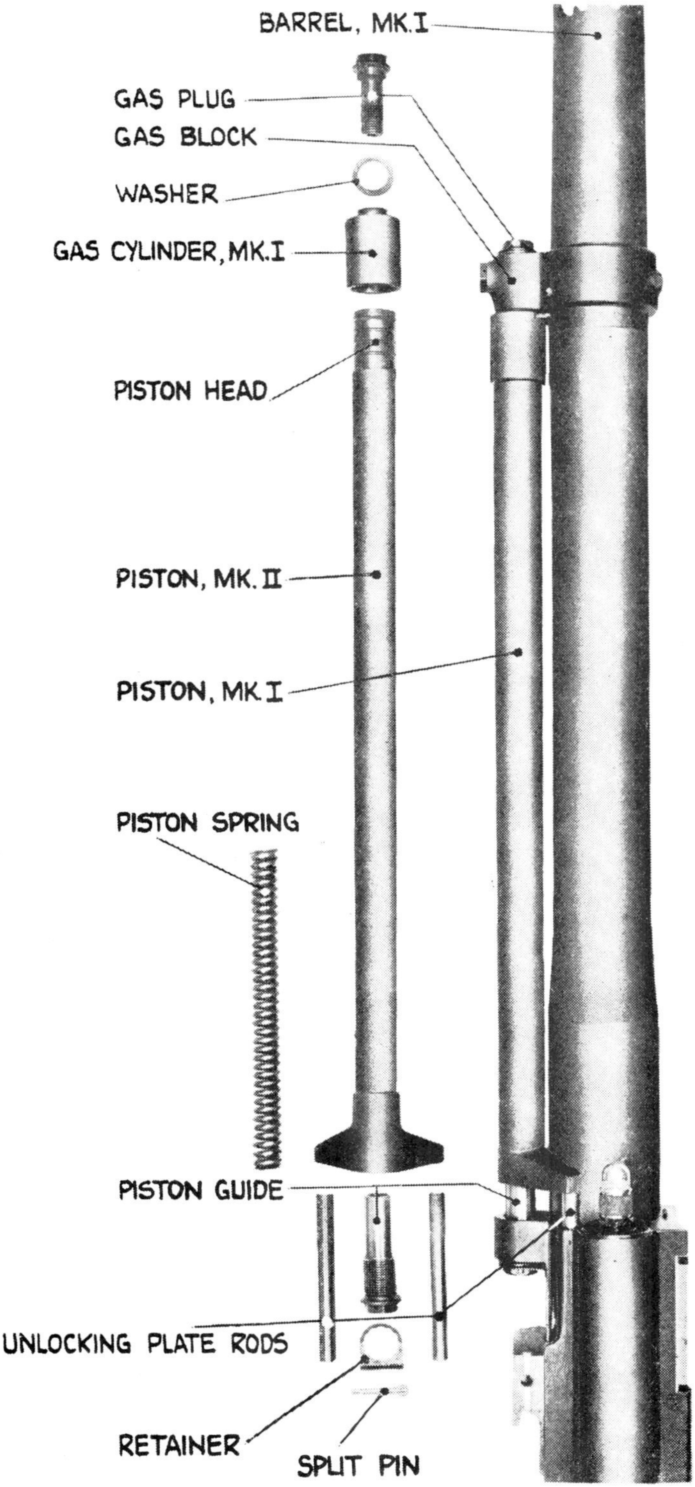

Fig. 7.—Gas cylinder and piston rod assembly

Unlocking plate rods, fig. 7

54. The unlocking plate rods are housed in the front of the body, their function being to transmit the movement of the piston to the unlocking plates when the piston is driven to the rear by the gas escaping into the gas cylinder.

Piston, Mk. II, fig. 7

55. The Mk. II piston differs from the Mk. I piston in external dimensions, being reduced in diameter for the greater part of its length, and in having the fork brazed on to the rod.

Breech block assembly

56. The breech block assembly consists of the breech block, the unlocking plates with cotter, unlocking plate springs and guides, locking-piece, firing pin, extractor, extractor spring and guides.

Breech block, Mk. I, fig. 3, 8 and 9

57. The Mk. I breech block is machined from the solid, being bored from the rear to accommodate the Mk. I firing pin and return spring. The top edges of the block are cut away to clear the horns of the ejector whilst the front face is recessed to accommodate the base of the cartridge case.

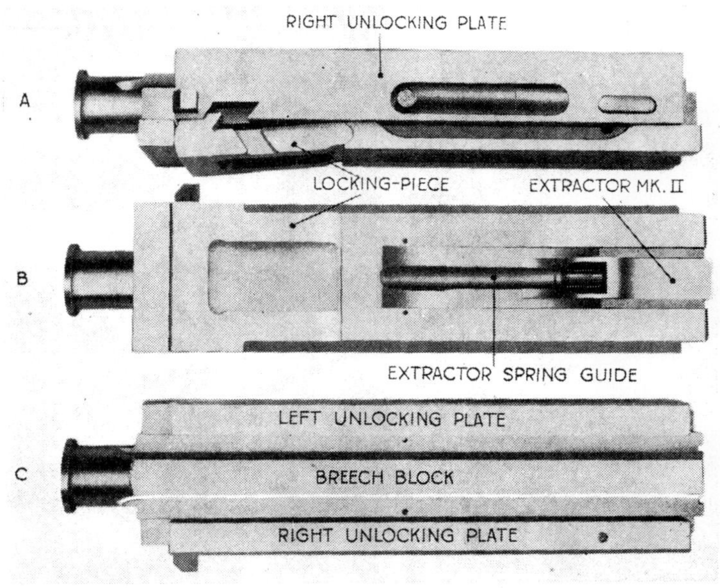

Fig. 8.—Breech block Mk. I

A—Right side view B—Bottom view C—Top view

58. The sides of the block and the underside at the rear are cut away to accommodate respectively the Mk. I unlocking plates and the locking-piece. Holes are drilled in the sides of the block

for the unlocking plates studs which are retained by pins. The studs are cylindrical in shape and form rear abutments for the unlocking plate springs. The underside of the breech block is recessed at the front for the Mk. II extractor and its spring and guides, a transverse hole being drilled through the block for the extractor pin. A transverse longitudinal slot is cut in the front of the block to accommodate the unlocking plates cotter.

Unlocking plates, Mk. I, fig. 8 and 9

59. The Mk. I unlocking plates fit against the sides of the Mk. I breech block and are connected at the front by the Mk. I cotter which is taper pinned to the right-hand plate and passes through the transverse slot in the breech block. The unlocking plates are identical except for a projection at the rear of the right-hand plate which is engaged by the cocking piston during the initial cocking of the gun. A longitudinal slot accommodates the unlocking plate spring which is housed in an outer and inner guide, the guides being telescopic. The outer guide bears against the stud in the breech block and the inner guide against the front of the slot in the unlocking plate. At the rear of the unlocking plate on the underside is a slot, the front and rear of which are in the form of ramps. The front or unlocking ramp engages a corresponding ramp on the locking-piece when the unlocking plate is forced to the rear during the opening of the breech. The rear ramp engages a bevel at the rear of the locking-piece and assists in lowering the locking-piece to the locked position. Behind the rear ramp is a flat which prevents the locking-piece from lifting when the breech is locked.

Locking-piece, fig. 3, 8, 9, and 10

60. The locking-piece is a flat steel plate, radiused at the front to engage in a correspondingly radiused groove in the underside of the breech block and form a hinge. Near the middle and on the sides, locking ramps are formed to engage ramps on the breech block guides when the breech block reaches the closed position. On the top and towards the rear, unlocking ramps are provided to engage corresponding ramps in the unlocking plates when the plates are forced to the rear. Behind the unlocking ramps on the locking-piece are flats which would foul the flats on the unlocking plates if the locking-piece were lifted before the plates had moved to the rear; this ensures positive engagement of the locking-piece with the locking shoulder at the moment of firing. At the rear, the bottom edge of the locking-piece is bevelled to enable it to hinge down and engage the locking shoulder, whilst the top edge is bevelled to form a ramp corresponding with the rear ramps on the unlocking plates. A bent is formed in the underside of the locking-piece to engage the sear bent.

61. The locking-piece is common to both Mk. I and Mk. II breech blocks.

Firing pin, Mk. I, fig. 9

62. The Mk. I firing pin is cylindrical in shape, being cut away to accommodate the unlocking plates cotter, which carries it forward to fire the gun when the breech is closed, and withdraws it inside the breech block as the unlocking plates move to the rear when unlocking the breech.

Extractor, Mk. II, fig. 8 and 9

63. The Mk. II extractor is accommodated in the underside of the breech block, and pivots on its pin. It is provided with a lip at the front to engage the groove at the base of the cartridge case. Near its pivot, the extractor is enlarged and recessed to accommodate the end of the inner guide for the extractor spring.

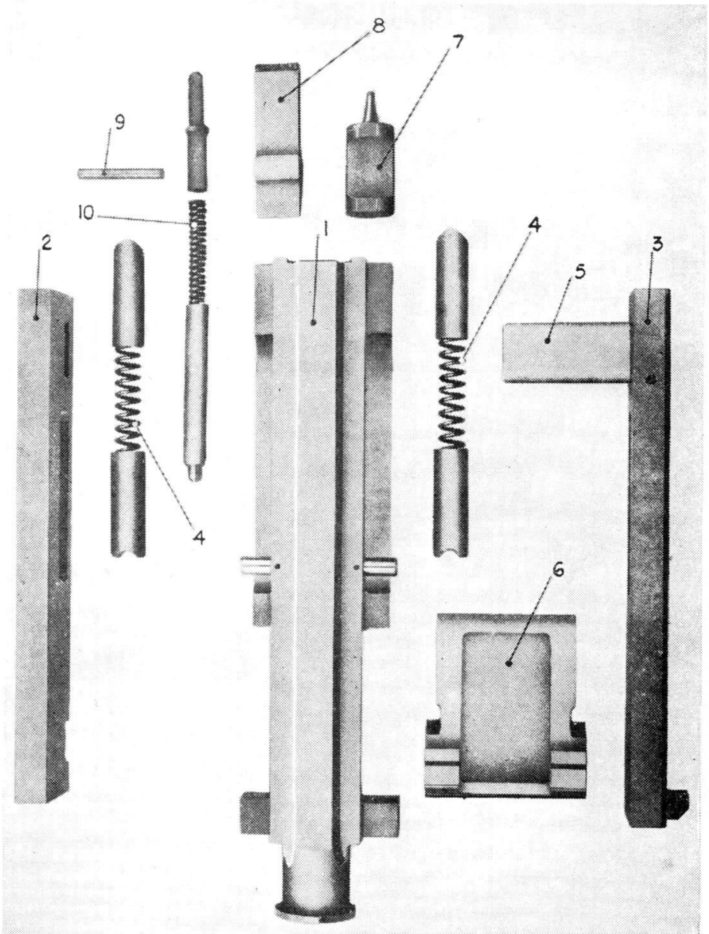

Fig. 9.—Breech block, Mk. I components

1. Breech block, Mk. I
2. Unlocking plate left, Mk. I
3. Unlocking plate right, Mk. I
4. Unlocking plate spring with outer and inner guides, Mk. I
5. Unlocking plates cotter, Mk. I
6. Locking-piece
7. Firing pin, Mk. I
8. Extractor, Mk. II
9. Extractor pin
10. Extractor spring with outer and inner guides, Mk. II

The end of the outer guide is accommodated in a recess towards the rear of the breech block. The extractor spring is housed inside the guides, which are telescopic, and its action is to push the lip of the extractor towards the face of the breech block.

64. The Mk. II extractor must not be fitted with the Mk. III ejector.

Breech block, Mk. II, fig. 10 and 11

65. The Mk. II breech block differs from the Mk. I breech block as follows:—

(i) The manufacturing tolerances are different.

(ii) The slot for the Mk. II unlocking plates cotter is narrower.

(iii) The studs for the Mk. II unlocking plate spring guides are rectangular in shape, and are drilled to accommodate the guides.

(iv) The block is fitted with Mk. II unlocking plates, Mk. II firing pin, and Mk. III extractor.

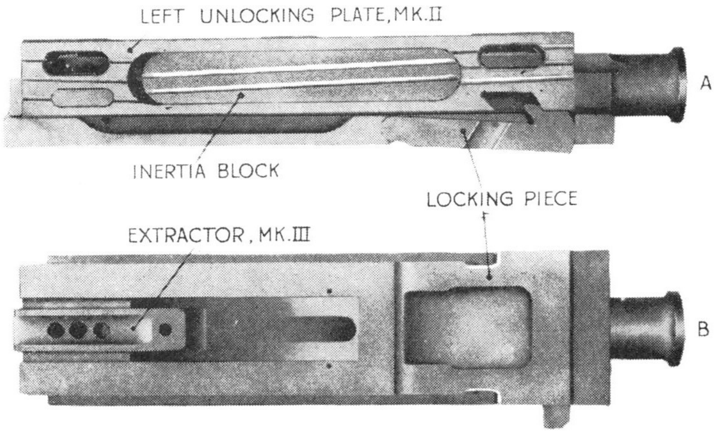

Fig. 10.—Breech block Mk. II

A—Left side view B—Bottom view

Unlocking plates, Mk. II, fig. 10 and 11

66. The Mk. II unlocking plates differ from the Mk. I unlocking plates as follows:—

(i) The Mk. II cotter is thinner than the Mk. I cotter.

(ii) Inertia blocks are provided and accommodated in the plates in the longitudinal slots, which are enlarged to suit them. The inertia blocks are cut away on the underside to accommodate the unlocking plate springs and guides, and are drilled at the front for a plunger and spring, the plunger bearing against the front of the slot in the unlocking plate. A recess in the plunger accommodates the pin which holds it in position and limits its movement. The head of the unlocking plate spring guide bears against the front of the slot in the unlocking plate, its opposite end passing through the stud in the Mk. II breech block. The unlocking plate spring is accommodated on the guide and bears between the head of the guide and the stud in the breech block.

(iii) The unlocking plates are lightened by a slot in each end.

(iv) Oil channels are provided.

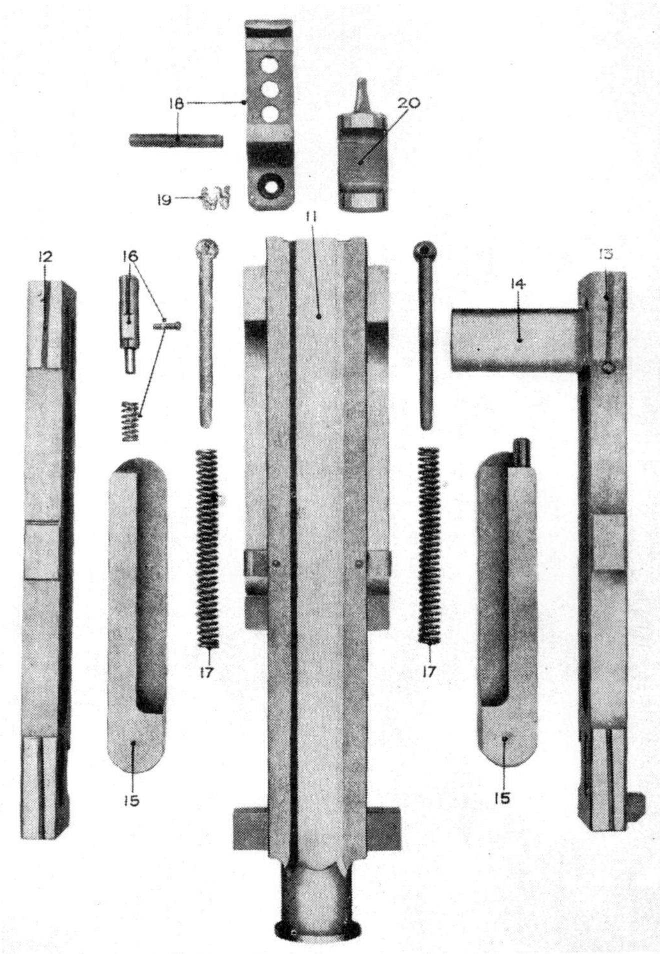

Fig. 11.—Breech block, Mk. II components

11. Breech block, Mk. II
12. Unlocking plate left, Mk. II
13. Unlocking plate right, Mk. II
14. Unlocking plates cotter, Mk. II
15. Inertia block
16. Inertia block plunger with spring and pin
17. Unlocking plate spring and guide, Mk. II
18. Extractor, Mk. III and pin
19. Extractor spring, Mk. III
20. Firing pin, Mk. II

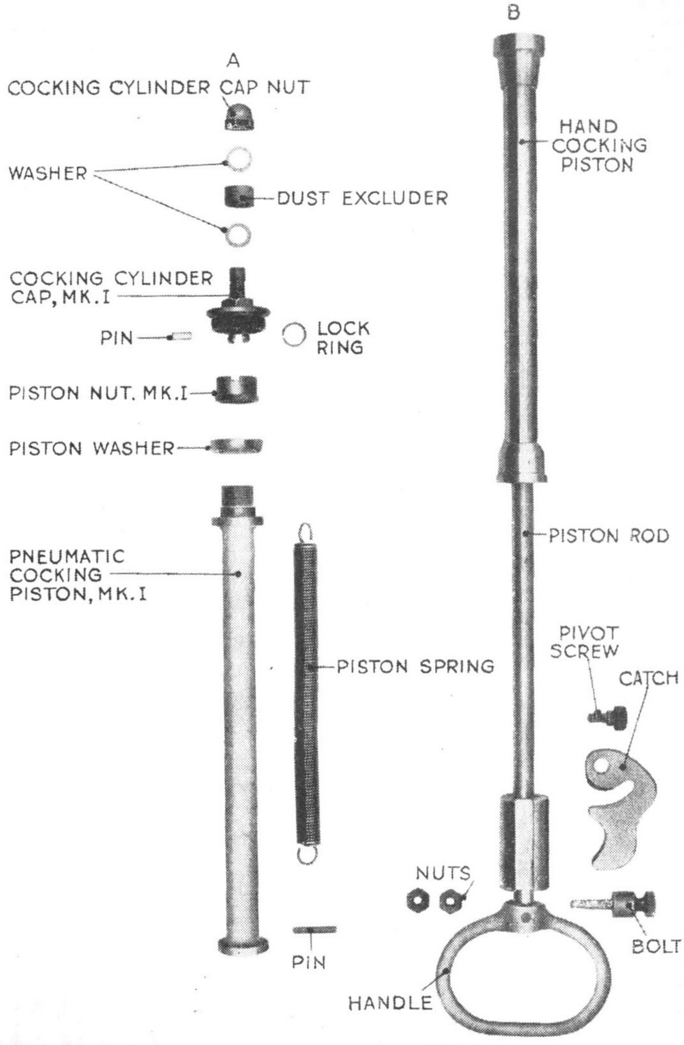

Fig. 12.—Pneumatic cocking unit, Mk. I and hand cocking unit

Firing pin, Mk. II, fig. 11

67. The Mk. II firing pin is similar to the Mk. I firing pin but has oil channels cut along its length.

Extractor, Mk. III, fig. 10 and 11

68. The Mk. III extractor differs from the Mk. II extractor in having a projection to the rear of the pivot and in being provided with a different type of spring. The spring is a short three-ply coil spring and is accommodated between the breech block and the projection on the extractor. The Mk. III extractor is lightened by three holes along its length.

Cocking units

Pneumatic cocking unit, Mk. I, fig. 12

69. The Mk. I pneumatic cocking unit consists of a Mk. I cocking piston with washer and piston nut, and Mk. I spring, and is accommodated in the cocking cylinder on the right of the gun body, which is closed by the Mk. I cocking cylinder cap.

70. The Mk. I cocking piston is hollow for almost the whole of its length and is flanged externally at both ends. The flange on the inner end engages the projection on the right-hand unlocking plate when the piston is forced to the rear by air pressure. Beyond the outer flange, the piston is threaded for the Mk. I piston nut which holds the piston washer in position against the flange. The Mk. I spring is housed inside the piston, its inner end being anchored by a pin which passes through the piston. The outer end of the spring is anchored to a pin which passes through a spigot on the inside of the Mk. I cocking cylinder cap, the ends of both pins are burred to prevent them from working loose.

Pneumatic cocking unit, Mk. II, fig. 13

71. The Mk. II pneumatic unit consists of a Mk. II piston with head, socket, washer, Mk. II nut, and nut retainer, Mk. II spring with securing bolt, washer, nut, retainer, and pin. It is retained in position in the cocking cylinder by the Mk. II cocking cylinder cap which is provided with a nut, washer, pin, and lock-ring.

72. The Mk. II cocking piston is a hollow cylinder having a head at its inner end, and a socket at its outer end. The Mk. II spring is accommodated inside the piston, its inner end being anchored to a pin which passes through the cocking piston spring securing bolt. The spring securing bolt is held in position in the piston head by a nut with split pin, a washer being placed between the head of the bolt and the inside of the piston head. The outer end of the spring is anchored to a pin which passes through a spigot on the inside of the Mk. II cocking cylinder cap, the pin being retained by a lock-ring. The socket has a shoulder formed on it, and is threaded for the Mk. II piston nut which holds the piston washer in position against the shoulder, the nut being secured by a retainer. A projection is formed on the outside of the cocking cylinder cap for the pneumatic union which is retained by a nut with copper washers. The cap is fitted with a dust excluder during transit.

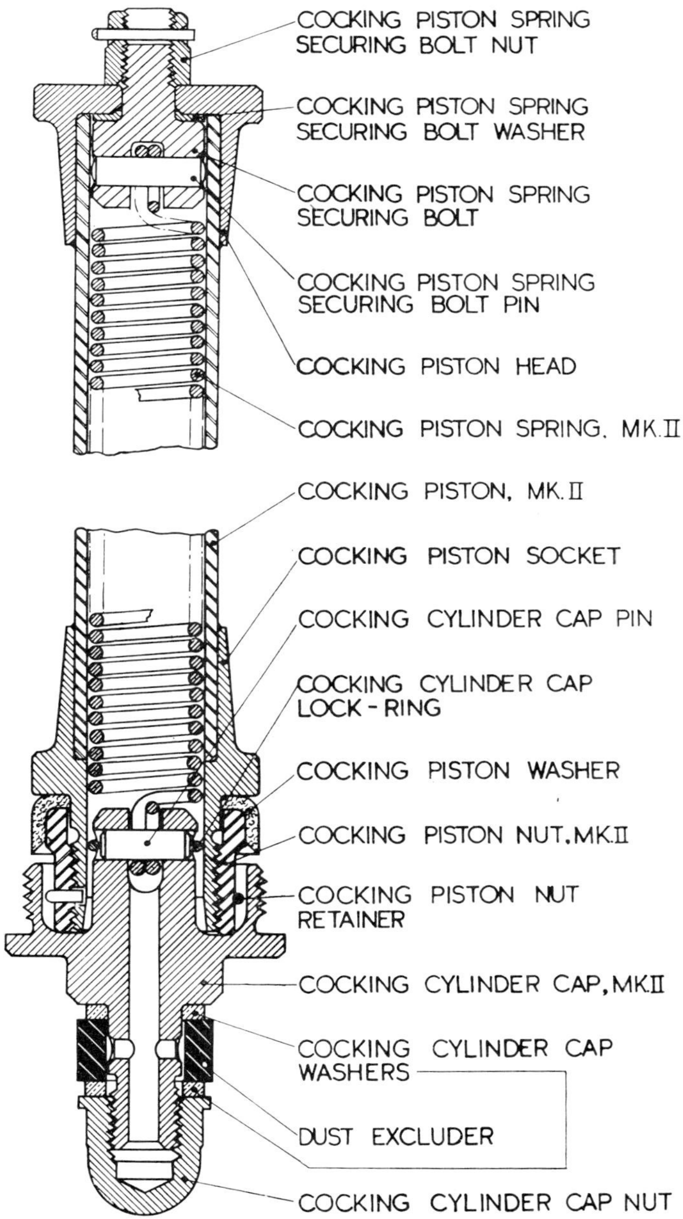

Fig. 13.—Pneumatic cocking unit, Mk. II, sectional diagram

Hand cocking unit, fig. 12

73. The hand cocking unit is for use on ground defence guns only. It consists of a piston with rod, handle, and catch, and is accommodated in the cocking cylinder. The cocking cylinder is open at the rear and is fitted with a bronze bush which is secured by an allen screw. The rod and piston are telescopic, the former projecting through the bush in the rear of the cylinder to which it is keyed. The end of the rod is threaded for the handle, which is prevented from unscrewing by a bolt, in the head of which is a circumferential groove for engagement of the catch. The catch is held in position by a pivot screw and has a notch cut in it to engage the handle securing bolt. After the gun has been cocked, the handle must be pushed forward and secured by the catch.

CHAPTER 3

Action of the gun

Initial cocking

74. When the cocking piston is forced to the rear, either by air pressure or by hand, the head of the piston engages against the projection on the right-hand unlocking plate, causing this and the left-hand unlocking plate, through the medium of the cotter, to move to the rear and compress the unlocking plate springs. The first movement of the cotter withdraws the firing pin inside the breech block, whilst the locking-piece is still in the locked position. When the unlocking plates have travelled approximately ½ in., the flats at the rear are clear of the flat at the rear of the locking-piece, thus allowing the locking-piece to be raised from the locked position through the engagement of the corresponding unlocking ramps on the unlocking plates and locking-piece. The breech block is now free to move to the rear, the cotter being at the rear of its slot.

75. Further movement of the cocking piston causes the unlocking plates cotter to pull the breech block to the rear, compressing the return spring between the firing pin and the rear of the return spring guide. The movement of the breech block is arrested when the projection at its rear hits the buffer washer. Towards the end of its movement, the breech block overrides the sear, which afterwards rises under the influence of its spring, and engages the locking-piece bent to hold the breech block in the cocked position.

76. The pneumatic cocking piston is returned to the forward position by its spring, which has been extended by the rearward movement of the piston, as soon as the air pressure is released. The handle and rod of the hand cocking unit are returned by hand, whilst the piston is returned by the subsequent forward movement of the unlocking plate.

Forward movement (assume a loaded magazine to be in position)

77. When the firing control mechanism is operated, the sear is disengaged and the return spring returns the breech block assembly to the forward position. The unlocking plates, cotter, and firing pin are prevented from moving independently of the breech block by the locking-piece. This is held against the breech block by the bottom plate, along which it rides, and the projecting ramps on the top of the locking-piece engaging the recesses in the unlocking plates prevent independent movement of the plates.

78. As the breech block moves forward, it meets the base of a round in the mouth of the magazine and pushes the round out of the magazine into the chamber. When the breech block is home, the extractor engages the groove in the cartridge case near the base.

79. The final movement of the breech block carries the locking-piece beyond the bottom plate and to the front of the locking shoulder. The locking-piece is lowered to the locked position by the locking ramps which engage the ramps of the breech block guides. The movement of the locking-piece frees the unlocking plates, which continue to move forward, and the rear of the ramps in the plates, riding over the bevel at the rear of the locking-piece, assist to lower this in front of the locking

shoulder. The locking-piece is prevented from lifting by the flats which engage the corresponding flats at the rear of the unlocking plates, thus positively locking the breech. The final movement of the unlocking plates carries the cotter to the front of the slot in the breech block. The firing pin, which is supported on the cotter, protrudes through the breech block face, and striking the cap of the cartridge, fires the gun. Rebound of the unlocking plates is prevented by their springs, which, in the Mk. II pattern, are assisted by the inertia blocks.

Backward movement (assume the gun to have just been fired)

80. The breech block remains positively locked until after the projectile has left the muzzle. When the projectile has passed the gas vent in the barrel, a portion of the propellant gases is deflected into the gas block and through the gas plug into the gas cylinder, where it impinges on the head of the piston and drives the piston to the rear. The rearward movement of the piston compresses the piston spring against the piston guide, the spring returning the piston to the front when the pressure is expended. As the piston moves to the rear, the projections on it engage and push the unlocking plate rods to the rear. The rods push the unlocking plates to the rear and the plates unlock the breech as described in the initial cocking, *see* para. 74.

81. When the breech is unlocked, the residual pressure in the bore acting against the empty cartridge case drives the breech block to the rear, compressing the return spring. The extractor supports the empty cartridge case on the face of the breech block until it meets the ejector which forces the case downwards through the ejector opening.

82. The backward movement of the breech block is arrested by the buffer in the back block. The breech block rebounds off the buffer and, if the firing control mechanism is operated, the cycle of events will be repeated, the return spring assisting the forward movement of the breech block. If the firing control mechanism is released, the sear will retain the breech block in the cocked position, and thus the gun does not normally stop firing with a live round in the chamber.

Action of the sear

83. If the firing control mechanism is released, the sear spring will reassert itself and cause the sear to project into the gun body. When the breech block travels to the rear, the locking-piece will override the sear, depressing it and compressing the sear spring. As soon as the locking-piece is clear, the sear will be returned by its spring ready to engage the locking-piece bent as the breech block rebounds off the buffer. When the locking-piece engages the sear, the latter is driven forward, causing the sear slide to compress the sear buffer springs against the plate and cushion to absorb the shock of engagement. At the same time, the ramp on the sear engages the sear cam in the trigger mechanism plate and ensures positive engagement of the bents. The sear buffer springs return the sear slide, sear, and breech block to their normal cocked position when the energy is expended. Operation of the firing control mechanism will disengage the sear from the locking-piece and allow the breech block to be driven forward by the return spring.

Action of the front mounting unit and recoil reducer

84. When the explosion of the charge takes place, the reaction of the gases on the cartridge case causes the gun to recoil. As the gun recoils, the front mounting unit sleeve compresses the spring against a shoulder in the front mounting cylinder, and, through the medium of the front mounting unit tube, forces the piston to the rear of the cylinder, into which air is drawn. After the energy of recoil has been absorbed by the Mk. I front mounting unit spring or the Mk. II double-rate recoil spring, the spring returns the gun to the firing position, and the piston, in expelling the air from the front mounting cylinder, cushions the run-out.

85. When the gun is fitted with a recoil reducer, the force of recoil is considerably reduced. As the projectile leaves the muzzle, the blast gases impinge on the collars in the choke, causing the gases in the recoil reducer cylinder to be deflected through the ports in the baffles and cylinder. The ports, being cut at an angle of 45 deg. deflect the gases to the rear, the tendency of the reaction being to force the gun to the front and reduce the total recoil energy.

Action of the magazine carrier

86. The magazine carrier is connected to the rear mounting by the magazine carrier arm, and therefore the magazine does not recoil.

Safety arrangements

87. The arrangement of the mechanism is such that until after the breech has been properly locked the gun cannot be fired for the following reasons:—

(i) If the breech block is not home, the locking-piece cannot be lowered as it would foul the locking shoulder.

(ii) The unlocking plates cannot move forward independently of the breech block unless the locking-piece is lowered into engagement with the locking shoulder.

(iii) Positive engagement of the locking-piece and locking shoulder is ensured by the rear of the unlocking plates riding over the flats on the locking-piece.

(iv) The firing pin, which is carried by the cotter, cannot protrude until the unlocking plates reach the end of their forward movement and cannot, therefore, strike the cap of the cartridge.

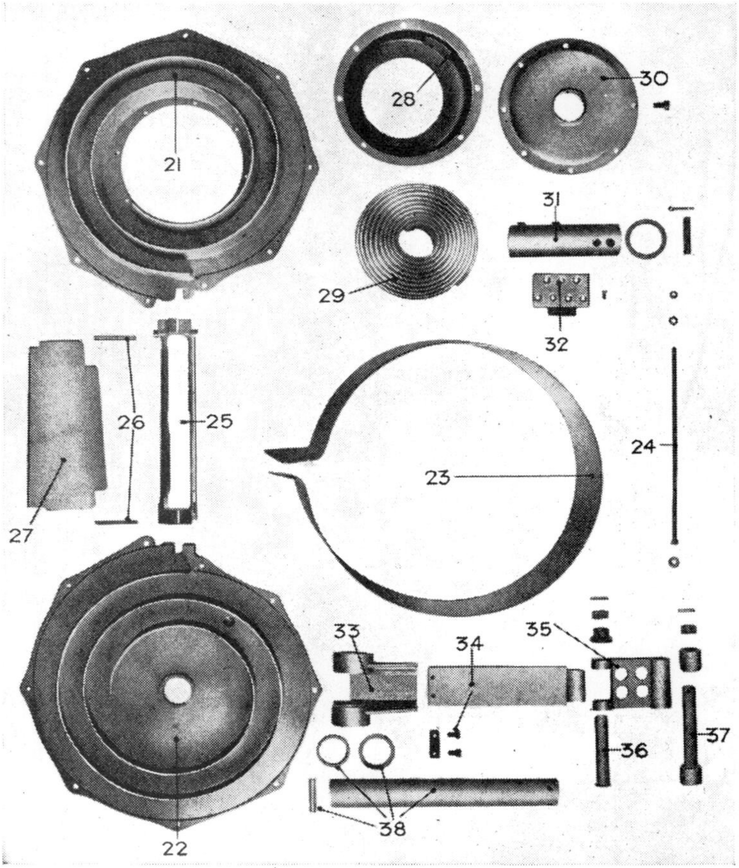

Fig. 14.—Magazine, No. 1, Mk. IIB components

21. Front plate
22. Rear plate
23. Outer casing
24. Tie rod with nut and spring washer
25. Mouth
26. Mouth pin
27. Throat plate
28. Spring casing
29. Spring
30. Front plate disk with screw
31. Tensioning tube with washer, pin, and split pin
32. Spring anchor plate with rivet
33. Feed arm bracket
34. Feed arm plate with stop and screws
35. Feed arm link
36. Feed arm axis bush with roller, collar, and taper pin
37. Platform tube with roller, collar, and taper pin
38. Feed arm axis tube with washer, collar, and pin

CHAPTER 4

Description of the HISPANO MAGAZINES

Magazine No. 1, Mk. IIB, fig. 14 and 15

88. The Hispano No. 1, Mk. IIB magazine is designed to hold 60 rounds of ammunition, and consists of the following main parts:—

(i) Front and rear plates with tie rods.

(ii) Outer casing.

(iii) Mouth with throat plate.

(iv) Spring and spring casing with front plate disk and tensioning tube.

(v) Feed arm axis tube with feed arm bracket, feed arm plate, link, axis bush, and platform tube.

Front and rear plates with tie rods

89. The front and rear plates are of cast steel and are held rigid against the outer casing by the tie rods, which are secured with single coil spring washers and nuts. On the inside of the plate are spirals which form guides for the ammunition. Tongues and grooves are formed on the edges of the plates for the atachment of the mouth. A hole in the rear plate accommodates the feed arm axis tube, whilst the spring casing is accommodated in the front plate. A window is provided in the rear plate to enable the ammunition to be inspected during loading.

Fig. 15.—Magazine in position on the gun

Outer casing

90. The outer casing consists of a band, the ends of which fit against the mouth and are held in position by two of the tie rods.

Mouth with throat plate

91. The mouth is a bronze casting with a tongue at each end to mate with those on the end plates, the mouth being secured to the end plates by mouth pins. Two pins are provided at the front of the mouth to engage the hook-shaped projections on the magazine carrier, whilst at the rear is a catch piece for engagement with the magazine catch when the magazine is in position on the gun. A throat plate is riveted to one side of the mouth to guide the ammunition into the spiral when loading the magazine, or into the mouth when loading the gun.

Spring and spring casing with front plate disk and tensioning tube

92. The outer end of the spring, which is of clock pattern, engages a spring anchor plate which is secured to the spring casing by seven rivets. The spring casing is held in position in the front plate of the magazine by the front plate disk, which is secured by seven screws. The inner end of the spring is anchored to the tensioning tube which fits over the end of the feed arm axis tube and projects through a bronze bush in the front plate disk. Holes are drilled in the outer end of the tensioning tube for the tensioning tube pin and for a tommy bar. Tension is applied to the spring by inserting a tommy bar in the holes and rotating it in a counter-clockwise direction. The pin which connects the tensioning tube to the feed arm axis tube is then inserted, thus transferring the tension to the latter tube.

Feed arm axis tube with feed arm bracket, feed arm plate, link, axis bush, and platform tube

93. The feed arm axis tube projects through the rear plate and is secured by a collar with pin and washer. The tube is drilled at the other end for the tensioning tube pin. The feed arm bracket is attached to the feed arm axis tube by two taper pins. Grooves are cut in the bracket to accommodate the feed arm plate which extends radially across the magazine. The plate slides in the bracket as the feed arm axis tube rotates, the radial movement of the plate being limited by a stop which is secured to it by two screws. At the end remote from the stop the feed arm plate is enlarged and drilled for the axis bush of the feed arm link which carries the platform or follower. One end of the axis bush carries a roller which is held in position by a collar secured by a taper pin. The roller follows the spiral as the feed arm rotates, and controls the movement of the feed arm plate in the bracket. One end of the platform tube carries a roller which is held in position by a collar secured by a taper pin. This roller follows the spiral as the feed arm rotates, and the platform or follower supports the last round in the magazine.

Magazine No. 1, Mk. IB

94. The No. 1, Mk. IB magazine differs from the No. 1, Mk. IIB magazine as follows:—

- (i) The front and rear plates are milled from the solid and have three grooves in each for the attachment of the mouth.
- (ii) The mouth has three tongues at each end.
- (iii) There is a slight difference in the length of the platform roller.

Magazine No. I, Mk. IA

95. The No. 1, Mk. IA magazine differs from the No. 1, Mk. IIB magazine as follows:—

(i) The front and rear plates are of pressed steel, and the spirals are spot welded in position,

(ii) The mouth has three tongues at each end, the front and rear plates being tongued and grooved to correspond.

Magazine No. 3, Mk. IA

95a. The No. 3, Mk. IA magazine differs from the No. 1, Mk. IA magazine in that it is fitted with a Mk. I impeller, see para. 96a to 96e, in the L.H. position on the magazine.

Magazine No. 4, Mk. IA

95b. The No. 4, Mk. IA magazine differs from the No. 1, Mk. IA magazine in that it is fitted with a Mk. I impeller, see para. 96a to 96e, in the R.H. position on the magazine.

Magazine No. 5, Mk. IA

95c. The No. 5, Mk. IA magazine differs from the No. 1, Mk. IA magazine in that it is fitted with a Mk. II impeller, see para. 96g to 96l, in the central position on the magazine.

Magazine No. 6, Mk. IA

95d. The No. 6, Mk. IA magazine differs from the No. 1, Mk. IA magazine in that it is fitted with a Mk. IA impeller, see para. 96f, in the L.H. position on the magazine.

Magazine No. 7, Mk. IA

95e. The No. 7, Mk. IA magazine differs from the No. 1, Mk. IA magazine in that it is fitted with a Mk. IA impeller, see para. 96f, in the R.H. position on the magazine.

Magazine No. 8, Mk. IA

95f. The No. 8, Mk. IA magazine differs from the No. 1, Mk. IA magazine in that it is fitted with a Mk. IA impeller, see para. 96f, in the central position on the magazine.

Magazine No. 9, Mk. IA

95g. The No. 9, Mk. IA magazine differs from the No. 1, Mk. IA magazine in that it is fitted with a Mk. I impeller, see para. 96a to 96e, in the central position on the magazine.

Action of the magazine

95h. Initial tension is applied to the spring during assembly, and, when the magazine is to be used, further tension is applied progressively during the loading operations, see para. 102 to 108.

95i. When the gun is being fired, the tensioned magazine spring, acting through the tensioning tube, feed arm axis tube, and feed arm, maintains the platform or follower in contact with the last round, and ensures that a round is always in position in the magazine mouth. As soon as one round is pushed out of the magazine mouth the next round is brought into position by the action of the spring. When an impeller is fitted, its blades engage the rounds and assist the magazine spring.

Hispano 20 mm. magazine impellers

96. The Hispano 20 mm. magazine impellers are fitted to improve the performance of the magazines. Each impeller comprises a spring-loaded shaft driving an impeller wheel having blades which project through an aperture in the magazine casing to give additional impetus to the rounds of ammunition and ensure that they are correctly positioned as they are fed into the magazine mouth. Instructions for loading and unloading magazines are given in para. 102 to 112.

Fig. 15a.—Top and bottom views of the Mk. I impeller

Impeller, Mk. I, fig. 15a and 15b

96a. The Mk. I impeller comprises a frame, cover, rear plate, driving spring, driving spring case and cover, shaft, driving spring hub, impeller wheel, and a number of smaller components.

Frame, cover, and rear plate

96b. The frame consists of a plate with integral flanges and lugs, and a transverse plate welded to it. A channel is attached to each side flange of the frame, and the semi-cylindrical cover has flanges which slide in these channels. The cover is retained in position by a small projection formed near the front edge of the cover engaging the transverse plate. The rear edge of the cover is flanged and helps to secure the rear plate, which is also located by the tie-rods when on the magazine.

Driving spring and driving spring case

96c. The driving spring is of clock pattern, with a hook at each end, and it is accommodated in a case which is secured by two studs to the transverse plate of the frame, its outer end hooking over one edge of a gap in the case rim. The rear of the case is closed by a flat cover carried on the shaft.

Shaft, driving spring hub, and impeller wheel

96d. The hollow shaft is mounted in a bush in the driving spring case and in the hub of the impeller wheel which rotates in a bush in the rear plate. On the front of the shaft is a collar secured by a split pin. Inside the driving spring the shaft carries a driving spring hub, which has a tooth to engage the hooked inner end of the spring and is prevented from rotating on the shaft by a pin passing transversely through the shaft and engaging a slot inside the hub.

96e. The impeller wheel, which has ten blades, fits loosely on the shaft, the latter projecting beyond the hub of the impeller wheel. A driving pin, which is secured to a retainer in the form of

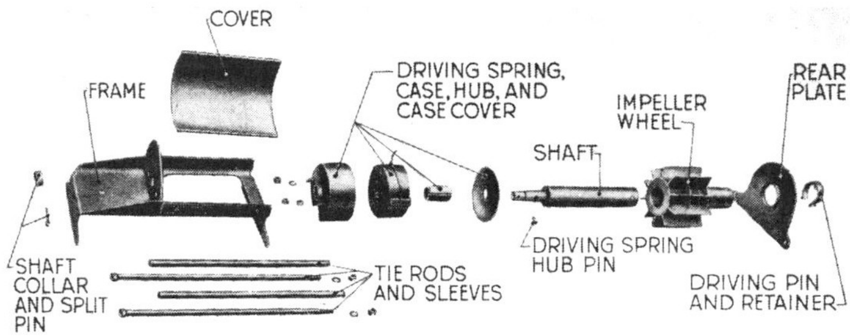

Fig. 15b.—Components of the Mk. I impeller

a clip which embraces the impeller wheel hub, passes through the impeller wheel hub and into the shaft to prevent the wheel from rotating relatively to the shaft. When it is desired to rotate the shaft without the wheel, during loading and unloading operations, the driving pin is removed. A tensioning bar is provided for winding up the impeller.

Impeller, Mk. IA

96f. The Mk. IA impeller is generally similar to the Mk. I impeller, but, instead of a driving pin on a retainer, it has a clutch collar and pin identical with those on the Mk. II impeller, see para. 96l.

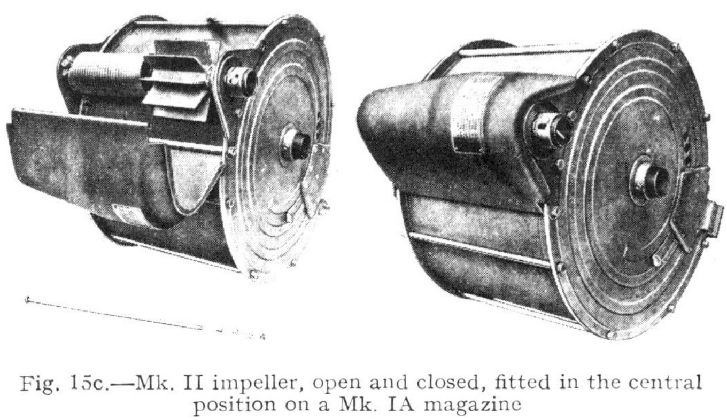

Fig. 15c.—Mk. II impeller, open and closed, fitted in the central position on a Mk. IA magazine

Impeller, Mk. II, fig. 15c and 15d

96g. The Mk. II impeller comprises a cover, front and rear plates, driving spring, shaft, impeller wheel, and a number of smaller components. A tensioning bar is provided for winding up the impeller.

Cover, and front and rear plates

96h. The cover is flanged internally at the ends and is secured to the magazine by two tie rods which also help to secure the front and rear plates. The front and rear plates are each additionally located by two bolts and by the flanges on the cover. Each plate has a bush, and on the front plate is a bracket with a stud to secure one end of the driving spring.

Driving spring

96j. The helical driving spring surrounds the shaft, its rear end being secured by a set-screw to the shaft, while the front end is secured to the front plate.

Shaft and impeller wheel

96k. The hollow shaft is mounted in the bushes in the front and rear plates, and has a collar near the front to engage the front plate, thereby locating the shaft axially. Near the centre of the shaft is a boss for the set-screw securing the driving spring to the shaft.

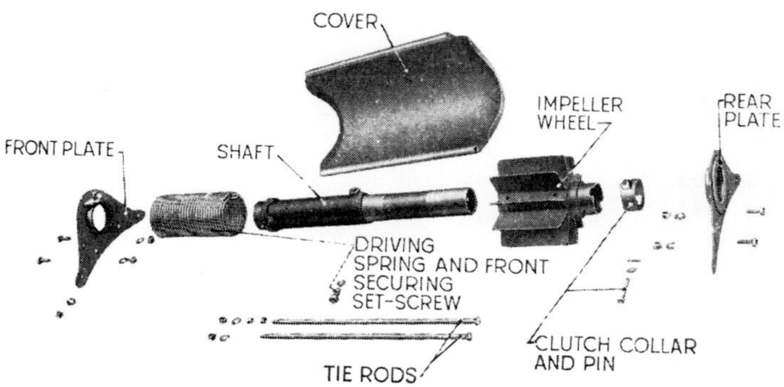

Fig. 15d.—Components of the Mk. II impeller

96l. The impeller wheel, which has ten blades, fits loosely on the shaft. A pin passing through the shaft locates a clutch collar with two dogs which engage corresponding slots in the hub of the impeller wheel to connect the wheel to the shaft. The clutch collar has L-shaped slots so that it can be turned into or out of engagement with the wheel hub; the torque of the spring normally holds the clutch collar in the engaged position.

Magazine No. 1, Mk. IA

95. The No. 1, Mk. IA magazine differs from the No. 1, Mk. IIB magazine as follows:—

(i) The front and rear plates are of pressed steel and the spirals are spot welded in position.

(ii) The mouth has three tongues at each end, the front and rear plates being grooved and tongued to correspond.

Action of the magazine

96. Initial tension is applied to the spring during assembly, and, when the magazine is to be used, further tension is applied progressively during the loading operations. The tensioned spring, acting through the tensioning tube, feed arm axis tube, and feed arm, maintains the platform or follower in contact with the last round, and ensures that a round is always in position in the magazine mouth. As soon as this round is loaded, the next round is brought into position by the spring.

CHAPTER 5

Installation, loading and unloading, operation, and stoppages

Installation

97. The gun may be fitted with one of the following combinations, depending upon the installation:—

(i) Recoil reducer, Mk. I or Mk. II, recoil reducer retainer Mk. I, and lock-ring Mk. I.

(ii) Front mounting unit No. 1, Mk. I, recoil reducer Mk. I or Mk. II, recoil reducer retainer Mk. II, and lock-ring Mk. II.

(iii) Front mounting unit No. 2, Mk. I, muzzle thread protector, muzzle thread protector retainer Mk. II and lock-ring Mk. II.

(iv) Muzzle thread protector, muzzle thread protector retainer Mk. I, and lock-ring Mk. I.

98. When the gun is correctly installed in its front and rear mountings and before the front mounting unit is fitted, it should be possible to move the gun freely and smoothly in a fore-and-aft direction. Some effort may be required to move the gun because of its weight, but if the movement is not free, malalinement in the front and rear mountings is indicated.

99. If the front mounting unit sleeve does not correctly engage the splines, the recoil reducer or muzzle thread protector cannot be screwed up to its correct position, and excessive recoil will be the result. Care should therefore be taken when mounting the gun to ensure that the splines are engaged, and that the sleeve abuts the front mounting unit tube.

100. To ensure that the magazine carrier is properly located relative to the gun, the magazine carrier arm must be adjusted until the lines on the magazine carrier and gun body coincide.

101. When the gun is mounted on its side, a support is provided for the magazine. This support must be adjusted so that the magazine is neither hanging on the gun nor tending to lift the gun. The mouth of the magazine should enter the magazine carrier easily, and not bear against it.

To load the magazine by hand

102. Do not oil or grease the ammunition or the mouth of the magazine.

103. If the magazine has been dismantled, initial tension must be applied as described in Vol. II, Part 3, Sect. 1, Chap. 2, para. 12.

104. Place the magazine on the magazine holder and, when loading the No. 1, Mk. IA or the No. 1, Mk. IB magazine, ensure that the distance piece is properly secured.

105. Pass the tommy bar through the holes in the tensioning tube and turn it slightly in a counter-clockwise direction to lower the platform or follower, and allow a round to be inserted.

106. Taking care not to knock the base of the cartridge case, insert a round base first into the mouth of the magazine and push it against the rear plate. Ease the tommy bar and ensure that the round is correctly positioned in the mouth.

107. Turn the tommy bar slightly in a counter-clockwise direction to lower this round, and insert the next round as described in para. 106.

108. Repeat the operation described in para. 107 until the magazine is full, taking care to maintain contact between the platform or follower and the first round inserted. No further tension must be applied to the spring.

Note.—Care must be exercised when loading the magazine, otherwise misfeeds may result.

To unload the magazine by hand

109. Place the magazine on the magazine holder, observing the precautions mentioned in para. 104.

110. Pass a tommy bar through the holes in the tensioning tube, and remove the pressure from the round in the mouth by rotating the tommy bar slightly in a counter-clockwise direction.

111. Remove the round from the mouth, taking care not to drop it.

112. To remove successive rounds, gradually ease the tommy bar, removing each round as it is brought into position in the mouth.

To load the gun

113. Cock the gun by operating the pneumatic cocking unit or hand cocking unit and set both the firing button and the pneumatic sear release unit to SAFE.

114. Take a full magazine and engage the two pins at the front of the mouth with the hook-shaped projections at the front of the magazine carrier. Operate the magazine catch lever and press the rear of the magazine into position.

115. Release the magazine catch lever and ensure that the catch engages the catch piece on the magazine.

116. Test the security of the magazine by endeavouring to remove it without operating the magazine catch lever. Ensure that the safety catch of the pneumatic sear release is set to FIRE before the aeroplane takes off.

To unload the gun

117. Ensure that the gun is pointing in a safe direction. Cock the gun and set the safety catch to SAFE. Operate the magazine catch lever and remove the magazine.

118. Fire the gun. Cock the gun and fire.

119. The gun is now unloaded.

Operation

120. When ready to fire, set the firing button to FIRE.

121. The pilot must not attempt to remedy a stoppage in the air by re-cocking the gun and pressing the firing button.

This may cause an explosion in the breech opening due to the cap of a live round in the chamber being struck by the nose of the next round, and serious damage would result. After landing, the pilot should taxi to the butts, or other safe place, to enable the gun to be examined.

Stoppages

122. The stoppages usually encountered fall into the following groups:—

(i) Misfeeds with the breech block jammed halfway, due to:—

(*a*) Bent round in the breech opening caused by insufficient tension on the magazine spring or friction due to dirt or thick oil.

(*b*) Damaged round in the breech opening due to the nose having missed the chamber and struck the end of the barrel. The round would probably be dislodged when the gun is cocked.

(*c*) Incorrect setting of the magazine carrier arm.

(ii) No feed with the breech block forward due to:—

(*a*) Empty or defective magazine.

(*b*) Incomplete backward action of the breech block caused by excessive friction, dirt or thick oil.

(iii) Misfire with the breech block forward, due to:—

(*a*) Unlocking plates not completing their forward movement as the result of dirt or thick oil between the unlocking plates and the breech block, or in the cotter slot.

(*b*) Defective cartridge cap.

(*c*) Cartridge rammed too far into the chamber due to an oversize chamber or undersize cartridge case.

(*d*) Insufficient protrusion of the firing pin.

(iv) Ejection failures with the breech block jammed in a partially open position, due to:—

(*a*) The empty case being caught between the breech block and the rear of the gun, which may be caused by a faulty extractor, or empty case deflector when the gun is mounted on its side.

(*b*) Broken ejector.

123. When a stoppage occurs, proceed as follows:—

(i) Note the position of the breech block.

(ii) Cock the gun and leave the air pressure on, so that there will no risk of a jammed breech block travelling forward when the air pressure is released.

(iii) Remove the magazine.

(iv) Examine the gun.

124. It is important that the magazine should not be removed before the gun has been cocked, or it may result in freeing a jammed gun and so firing a round accidentally.

CHAPTER 6

Cleaning rod and tools

Cleaning rod, fig. 16

125. The cleaning rod consists of a number of metal sections, having male and female threads at opposite ends to enable the complete rod to be assembled. A handle, which may be in the

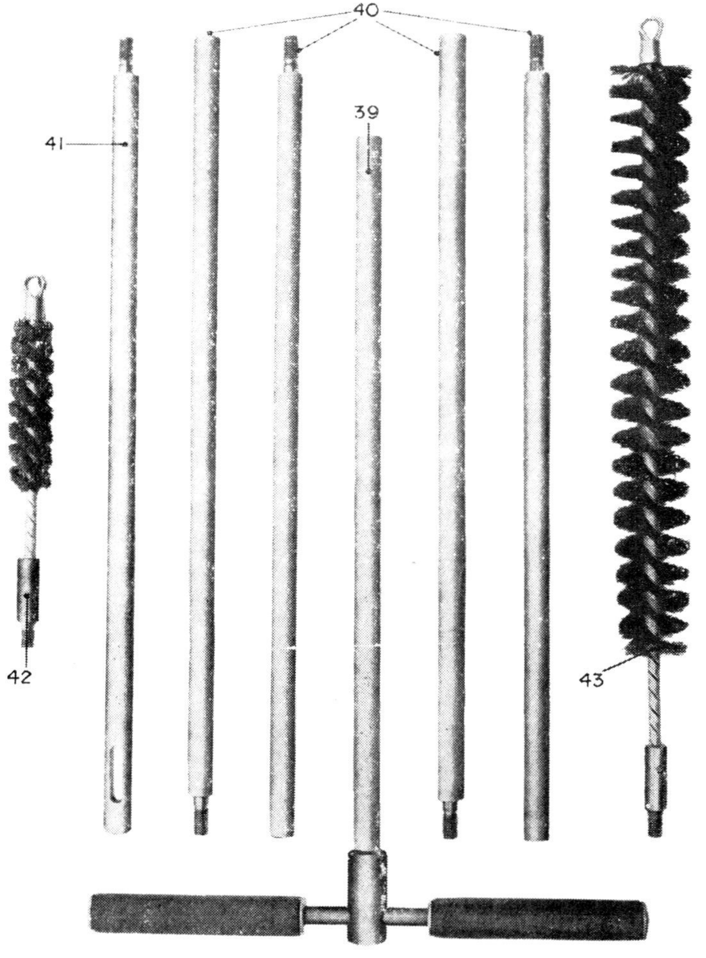

Fig. 16.—Cleaning rod

39. Handle
40. Sections
41. Loop
42. Wire brush
43. Bristle brush

form of a ring or a cross piece, is fitted to one of the sections to facilitate use of the rod. A slot is cut in another section to form a loop for flannelette or other cleaning rag. A bristle brush and a wire brush are provided for use with the rod, each being threaded at one end to screw into the end section.

Tools, fig. 17

126. The return spring assembling tool complete with spanner was provided for use with the Mk. I return spring and Mk. I return spring guide. It consists of a steel rod with a wood handle, and when in use the rod is passed through a hole in the centre of the spanner, which is fitted with cross-pieces.

127. The buffer spring sleeve key is a cylindrical tube with an external flange at one end in which are cut keyways to mate with the keys in the sleeve, when assembling and dismantling. At the opposite end cross-pieces are fitted.

128. The recoil reducer assembling tool is a casting with two handles and having internal splines to engage the splines on the recoil reducer when assembling and dismantling. It can also be used for assembling the muzzle thread protector when this is in use.

129. The return spring guide spanner is for use with the Mk. II return spring guide. It can be distinguished from the spanner used with the Mk. I return spring guide as it has keyways cut in one face to mate with the keys in the buffer spring sleeves, and may be used for assembling and dismantling this part.

130. The return spring guide rod assembling tool is a steel rod in one end of which is screwed a split stud. The stud is inserted in the outer end of the Mk. II return spring guide rod when assembling and dismantling.

131. The sear buffer springs clamp is designed to enable the sear buffer springs to be compressed or to enable the pressure on the springs to be released. A hole is drilled in the end opposite to the screw clamp to permit the assembling tool to be inserted.

132. The sear buffer springs assembling tool is used in conjunction with the clamp, and when inserted in the hole in the sear slide engages the grooves in the sear buffer springs plugs, to hold the springs under compression; thus enabling the sear slide to be removed from the clamp and assembled in the gun, or removed from the gun and placed in the clamp.

133. The tool for unlocking the breech and compressing the magazine catch springs consists of a lever having two arms pivoted at one end. It is used for unlocking the breech when dismantling, and for compressing the Mk. I magazine catch springs when dismantling and assembling.

134. The magazine holder is suitably shaped to accommodate the No. 1, Mk. IIB magazine, an adapter being provided for use with the No. 1, Mk. IA and No. 1, Mk. IB magazines. The adapter is secured in the holder by wing nuts. Holes are drilled in the bottom of the holder for the woodscrews which secure it to the bench, and in the rear are holes to accommodate the end of the feed arm axis tube and one of the tie rods.

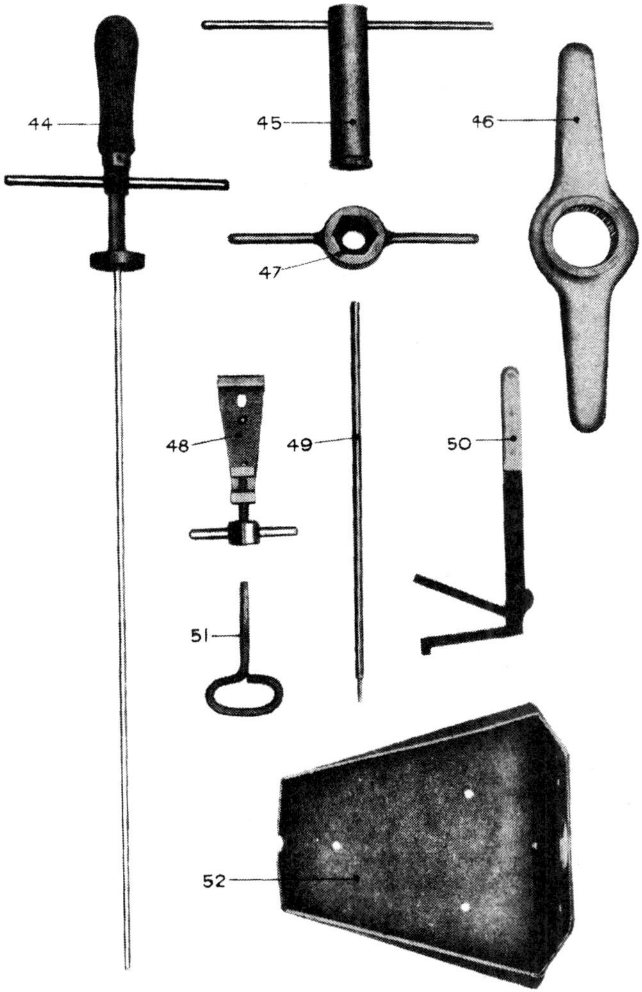

Fig. 17.—Tools

44. Return spring assembling tool with spanner
45. Buffer spring sleeve key
46. Recoil reducer assembling tool
47. Return spring guide spanner
48. Sear buffer springs clamp
49. Return spring guide rod assembling tool
50. Tool for unlocking breech and compressing magazine catch springs
51. Sear buffer springs assembling tool
52. Magazine holder

CHAPTER 7

Description and action of, loading and unloading, the HISPANO 20 mm. BELT FEED MECHANISM, Mk. I (Left-hand and right-hand)

Introduction

135. The belt feed mechanism, which is spring-operated, is an alternative feed to the 30 or 60-round magazines. The rounds are assembled with links into a belt and the function of the belt feed mechanism is to draw up the belt from the ammunition container, separate the rounds from the links, and feed the rounds one at a time into the breech of the gun. The belt feed mechanism, which is mounted on the magazine carrier, can be fitted to any Hispano 20 mm. Mk. I or II gun, providing a Mk. II or III magazine carrier is used, and is made for either right-hand or left-hand feed.

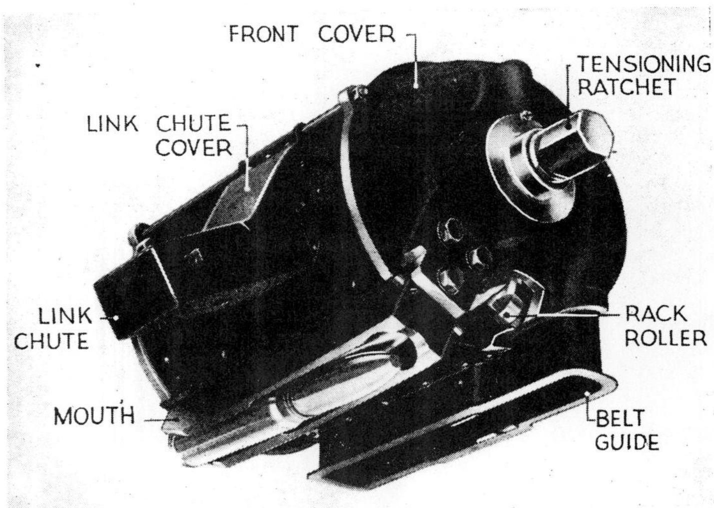

Fig. 18.—Perspective view of Hispano 20 mm. belt feed mechanism, .Mk. I, for L H. feed

136. The belt is made up with links having loops which are open at one side to permit disintegration of the belt after the rounds have been withdrawn from the links through a distance equal approximately to the width of a link. The belts must be assembled for either right-hand or left-hand feed as required, the shape of the links being such that they can be used for making up either form of belt. To prevent the last link from becomiug caught up inside the feed mechanism, its single loop is closed by a bridge or strip secured across the open side of the loop. Special links with a closed loop of this kind are provided for use when the belts are made up.

General description, fig. 18 and 19

137. The belt feed mechanism comprises a cylindrical metal case, the ends of which are closed by front and rear covers held by three tie rods. Inside the case is a rotatable central shaft, the

ends of which are supported by the covers. Three sprockets, with hubs keyed on the shaft, form an assembly which rotates as a whole. A link ejector bracket is mounted on the hub of the front sprocket, a front feed lever, carrying a last round retainer, is mounted on the hub of the centre sprocket, and a rear feed lever is mounted on the hub of the rear sprocket. Riveted to the front sprocket is a driving spring case, inside which is a spiral driving spring which is wound up initially to its maximum tension, a spring band brake being provided to obviate the risk of overwinding the spring. As each round is fired the recoil of the gun actuates a rack mounted in the front cover. The rack engages a ratchet actuating segment which rotates a tensioning ratchet to maintain the tension of the driving spring. A pawl in the front cover engages teeth formed on the periphery of the tensioning ratchet to prevent unwinding.

138. The belt enters the belt feed mechanism through a belt guide, and the rounds are engaged by the sprockets and carried round between the sprockets and circumferential guides which are riveted inside the case. The noses of the bullets contact the bottom of a cam groove formed inside the front cover and the rounds are forced axially towards the rear. The links are prevented from moving towards the rear by the teeth of the centre sprocket, so that, when the rounds have been pushed back far enough, the belt disintegrates. The links are pushed clear of the rounds by a link ejector and fall through a link chute.

139. The rounds, after being freed from the links are conveyed past the feed levers and are deflected by an extension of the link ejector bracket into a mouth, which is similar to that of the magazine. The last round retainer, mounted on the front feed lever, projects into the mouth, and, while there is still a round between the sprocket teeth pressing on the feed levers, the retainer is not operative. When only the last two rounds remain in the mouth, the retainer prevents the last round from reaching the loading position.

140. For the purpose of description the mechanism is divided into the following groups:—

- (i) Case with belt guide, link chute, mouth, and front and rear covers,
- (ii) Driving mechanism,
- (iii) Shaft assembly,
- (iv) Rack-operating lever and bracket.

Case with belt guide, link chute, mouth, and front and rear covers

Case, fig. 19

141. The case is cylindrical, open at the ends and down one side. Inside the case are riveted three circumferential guides. The front and centre guides extend round nearly the full circumference of the case, but the rear guide is short, being located close to the mouth on the side nearer the link chute.

Belt guide, fig. 18

142. The belt guide is riveted over a slot in the case and its lower edge permits an ammunition chute to be attached. Two packing pieces, which guide the links between the sprockets, are riveted inside the belt guide; the thinner one on the side near the mouth, and the thicker one on the side away from the mouth.

Link chute, fig. 18

143. The link chute is riveted over a slot in the case and a cover is provided on top to give access to the interior of the chute and the belt feed mechanism. The cover is hinged to one of the tie rods, being prevented from lateral displacement by side flanges. A latch is riveted to the cover and retains it in the closed position by engaging a stud on the outer end of the chute. A ramp is secured inside the cover, and a flat spring, riveted to the inner end of the ramp, facilitates the ejection of the links. A lug welded to each side of the link chute permits attachment of a neck or extension to convey the links away from the gun.

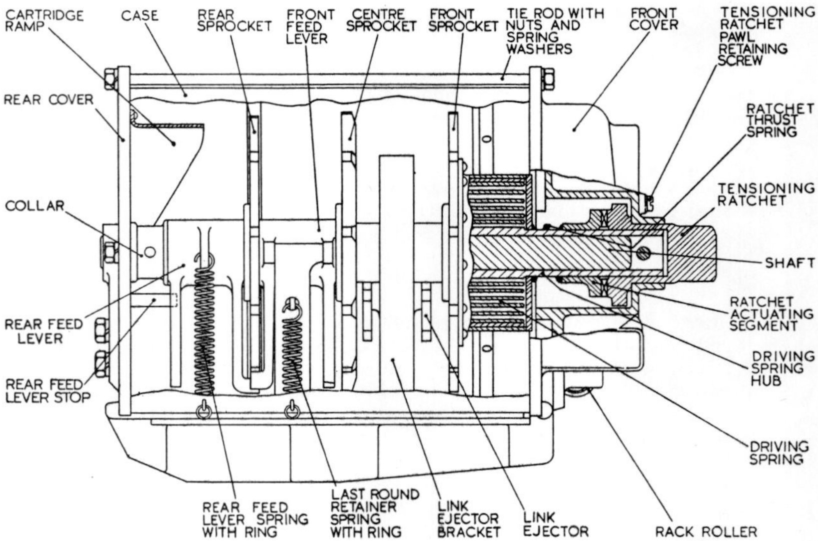

Fig. 19.—Part sectional view of Hispano 20 mm. belt feed mechanism, Mk. I, for R.H. feed

Mouth, fig. 18 and 20

144. The mouth is similar in its external part to the mouth of the 60-round magazine. A vertical lug at each end projects upwards inside the front and rear covers, which are recessed to accommodate them. Three tapped holes are provided in each lug, and the mouth is secured at each end by three screws which pass through the cover. The screws at the rear end are shorter than those at the front end. Grooves are formed along each side of the mouth to receive the edges of the case and at the rear of the mouth is a shoulder for the magazine catch to engage. On each side, at the front of the mouth, is a pin to engage the hook-shaped projections of the magazine carrier, and just forward of the centre, inside the upper edges, are formed two small recesses, one of which, according to whether the mouth is used on a right-hand or left-hand feed mechanism, accommodates one part of the forked end of the link ejector bracket which rests on the edge of the mouth. Two other recesses, one in each side, are formed inside the mouth behind the recesses for the link ejector bracket. At the lower end of each of these recesses is formed an inclined face, against one of which the lower end of the last round retainer lies according to whether the mouth is used on a right or left-hand feed mechanism.

A small lip is formed along part of the lower edge of the mouth on each side to support each round until it is pushed forward by the breech block.

Front and rear covers, fig. 20

145. The front and rear covers have lugs for the tie rods which hold them to the case, and each has a circumferential groove in its inner face to receive one end of the case. In the front cover is formed the cam groove by which the noses of the bullets are guided and thrust back as the rounds are carried through the mechanism. A hollow boss on the front cover accommodates the driving spring mechanism, and recesses are provided for the actuating rack and pawl. The rear cover is practically flat, but has a bored central boss which forms a bearing for the rear end of the shaft. A cartridge ramp, of the complementary contour to the cam groove in the front cover, is riveted to the rear cover and limits the rearward movement of the rounds. A rear feed lever stop, in the form of a peg, is also provided.

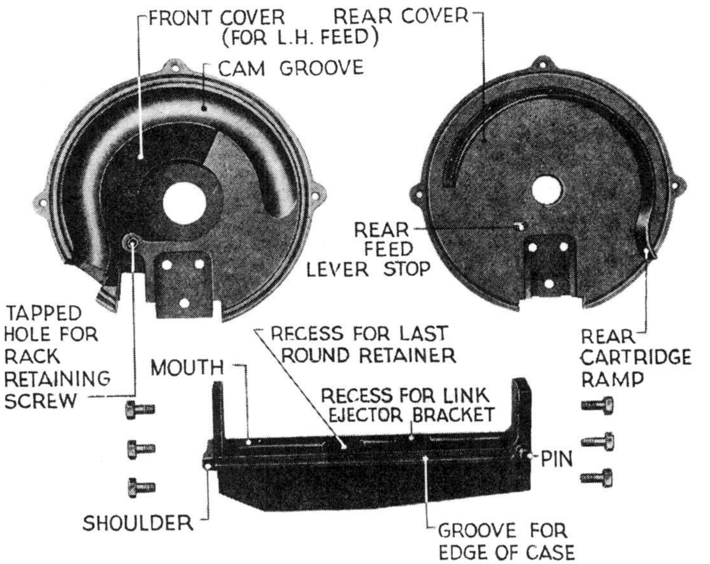

Fig. 20.—Covers and mouth

Driving mechanism, fig. 21

146. The driving mechanism consists of a rack with a roller spring and spring guide, a ratchet actuating segment, a tensioning ratchet with spring-pressed pawl, a ratchet thrust spring, a driving spring hub, a driving spring, and a driving spring band brake.

Rack, fig. 21

147. The rack has six involute teeth to engage the teeth on the periphery of the actuating segment. A blind hole is drilled in the upper end to accommodate the rack spring with its guide. At the lower end of the rack is a roller which rotates on a pin, the

ends of which are riveted over. Two longitudinal slots are provided in the lower half of the rack, diametrically opposite to each other, and one of these slots, according to whether the rack is used in a right-hand or left-hand feed mechanism, is engaged by the plain portion of the rack retaining screw.

Actuating segment, fig. 21

148. The actuating segment is a running fit on the driving spring hub, and is accommodated inside the hollow boss of the front cover. Eight involute teeth are formed on part of its periphery to engage with the teeth of the rack. On the front face of the segment 11 ratchet teeth are provided, to engage with the ratchet teeth of the tensioning ratchet. A ratchet thrust spring, one end of which bears against the driving spring case cover, urges the actuating segment against the tensioning ratchet to maintain contact between the two sets of teeth.

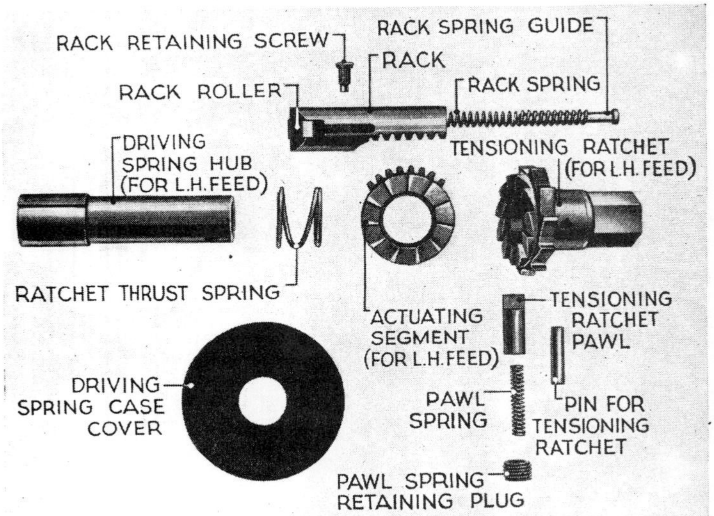

Fig. 21.—Driving mechanism components

Tensioning ratchet and pawl, fig. 21

149. The tensioning ratchet fits over the front end of the driving spring hub, to which it is secured by a riveted pin. The driving spring hub fits over and is free to rotate on the front end of the shaft, and the boss of the tensioning ratchet rotates in the bearing formed in the front cover. The tensioning ratchet thus forms the front bearing for the shaft, but each is free to rotate independently of the other. The rear face of the tensioning ratchet is provided with the 11 ratchet teeth which engage the teeth of the actuating segment. On the periphery of the tensioning ratchet are the 11 teeth which engage the pawl. The boss of the tensioning ratchet projects through the front cover and is hexagonal to permit a spanner or ratchet wrench to be used for applying the initial tension to the driving spring.

150. One of the flat faces of the pawl is extended partly along the side and engages the plain portion of the shank of the retaining screw, thus preventing the pawl from rotating while allowing it

to reciprocate. The upper end of the pawl accommodates one end of the pawl spring, the other end of the spring abutting a screwed plug fitted into the top of the pawl recess.

Driving spring hub, fig. 21

151. The driving spring hub fits over the front end of the shaft. At the rear end its diameter is increased and four ratchet type teeth are formed on the periphery, any one of which can engage the inner end of the driving spring. The front end of the hub fits inside the tensioning ratchet, to which it is secured by the pin passing diametrically through both hub and ratchet. The hub also carries the driving spring case cover, the ratchet thrust spring, and the actuating segment.

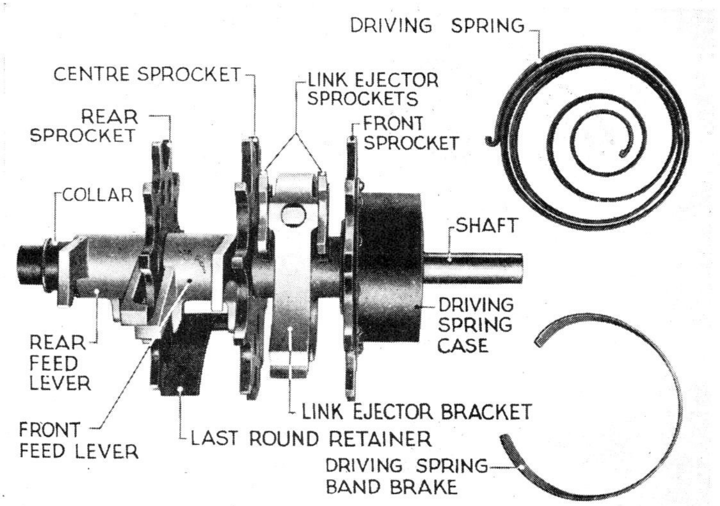

Fig. 22.—Shaft assembly, for L.H. feed

Driving spring and band brake, fig. 22

152. The driving spring is a flat coiled spring of clock type. At the inner end it is doubled back against itself for a short distance and engages any one of the four teeth on the driving spring hub. The outer end is formed into a hook to engage one end of the driving spring band brake.

153. The driving spring band brake is of the fixed pressure internally expanding type and is fitted inside the driving spring case in which it slips round when the spring has reached the maximum tension.

Shaft assembly, fig. 22

154. The shaft assembly consists of the shaft, front, centre, and rear sprockets, a link ejector and bracket, a front feed lever with last round retainer, a rear feed lever, and a collar.

Shaft, fig. 22

155. The shaft is solid and has two journals, that at the front being reduced in diameter to enter the driving spring hub, and

that at the rear being short and bearing directly in the rear cover. A key way and key are provided in the shaft to locate the sprockets, and holes are drilled for the collar and front sprocket pins.

Front, centre, and rear sprockets, fig. 19 and 22

156. The front, centre, and rear sprockets are provided with hubs which extend from their rear faces, each being provided with a key way to accommodate the shaft key. The hubs serve as spacers to maintain the sprockets in their correct longitudinal positions on the shaft. Each sprocket has 11 teeth and a ring of lightening holes. The driving spring case is riveted to the front face of the front sprocket. The centre sprocket is thicker than the others and prevents axial movement of the links while the rounds are pushed out of them towards the rear, and the front edges of the teeth are bevelled so that they do not foul the links and jam the mechanism.

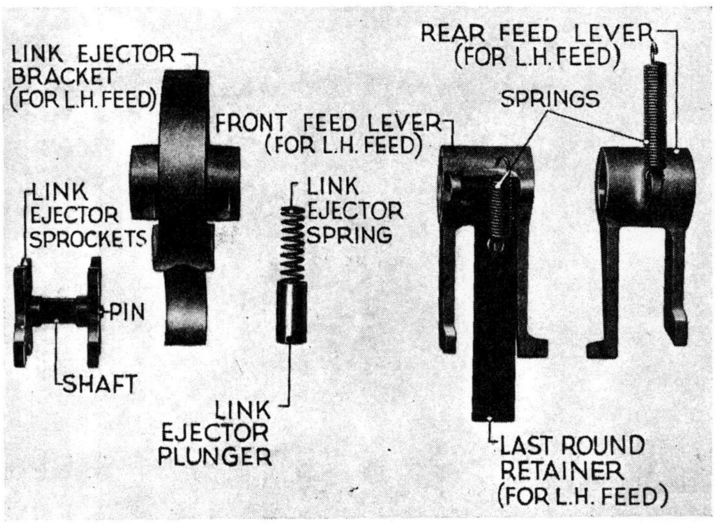

Fig. 23.—Feed levers and link ejector

Link ejector and bracket, fig. 22 and 23

157. The link ejector comprises two four-toothed sprockets rotatably mounted on a bracket which is maintained stationary between the front and centre main sprockets. A shaft is formed integrally with one of the ejector sprockets and has four flats on its centre portion. The other ejector sprocket is secured by a riveted pin passing longitudinally through the shaft. A spring-pressed plunger is located in the bracket to bear against the flats of the shaft and hold the link ejector in the correct position to be engaged by each round. The link ejector bracket is mounted on the hub of the front feed sprocket, which is free to rotate within the bracket. To prevent the bracket rotating, an arm, forked at its outer end, engages the top edge of the mouth on the belt guide side. The upper side of the arm is curved and helps to guide the rounds into the mouth.

Front feed lever with last round retainer, fig. 19, 22, and 23

158. The front feed lever has two arms and is mounted on the hub of the centre sprocket. The front arm has a projection to push on the upper side of the second round in the mouth, and the rear arm has an extension which projects towards the rear and mates with a projection on the rear feed lever. Normally the two feed levers operate as one unit, but one can turn relatively to the other in one direction, thus preventing damage if a round should enter the mouth obliquely. Two lugs are formed on the front feed lever, the last round retainer being pivotally mounted between them.

159. The last round retainer is a flat lever which projects into the mouth. On its outer side is a stud, and the last round retainer spring, which is a tension spring, is secured at one end to the stud. The other end of the spring is formed into a loop which is passed through a slot in the case between the mouth and the belt guide and anchored by a split ring. The spring holds the retainer against the belt guide side of the mouth and tends to pull the retainer further into the mouth.

Rear feed lever and collar, fig. 19, 22, and 23

160. The rear feed lever has two arms and is mounted on the hub of the rear sprocket. The rear arm is the shorter of the two and has a projection at its lower end to steady the rounds in the mouth. A lug is formed on the rear feed lever as an anchorage for one end of the rear feed lever spring, the other end of the spring being passed through a slot in the case, between the belt guide and the mouth, and secured by a split ring. The spring tends to rotate the lever in the opposite direction to that in which the shaft rotates, but the rear feed lever stop, inside the rear cover, engages the rear arm of the lever to prevent it rotating too far.

161. The collar is secured to the shaft by a pin and bears against the rear cover, thus locating the shaft axially, and keeping the rear feed lever and the centre and rear sprockets in their correct positions on the shaft.

Rack-operating lever, bracket, and piston guide, fig. 24

162. A bracket with a rack-operating lever is mounted with a No. 2 piston guide on the gun so that the recoil is used to actuate the rack.

Bracket and rack-operating lever, fig. 24

163. The bracket is mounted on the gun behind the piston guide boss, being retained by the No. 2 piston guide, see para. 165, and has its main portion overhanging the gun on the side from which the belt is being fed. The thinner portion of the bracket is located behind the piston guide boss and has a hole for the piston guide, which is locked by a taper pin. A slot is cut in the main portion of the bracket to accommodate the lever.

164. The rack-operating lever pivots in the bracket and at the rear end has an inclined face which engages the roller of the rack. In the lever, below the inclined face, is a roller, which engages the inclined face on the front edge of the Mk. II or III magazine carrier. When the gun recoils, the lever roller rides up the inclined face of the magazine carrier and the rack roller rides up the inclined face of the lever, the combined movement actuating the tensioning ratchet and maintaining the spring tension.

No. 2 piston guide

165. The No. 2 piston guide replaces the No. 1 piston guide when the belt feed mechanism is used. It has a plain stem of the same length as that of the No. 1 piston guide, but its screw-threaded portion is longer, and, instead of a hexagon head, it has an Allen head. On the head 12 flats are formed, and one of these is engaged by the taper pin of the bracket.

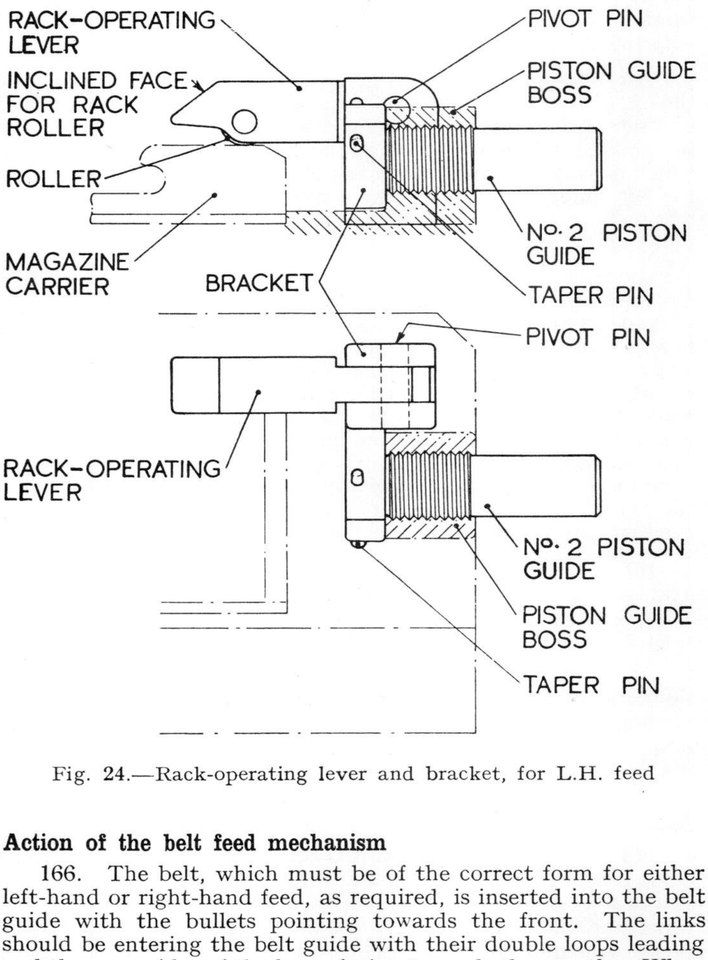

Fig. 24.—Rack-operating lever and bracket, for L.H. feed

Action of the belt feed mechanism

166. The belt, which must be of the correct form for either left-hand or right-hand feed, as required, is inserted into the belt guide with the bullets pointing towards the front. The links should be entering the belt guide with their double loops leading and the open sides of the loops facing towards the mouth. When the belt is being fed by hand into the mechanism the rounds engage the feed sprockets and cause the shaft, driving spring and driving spring case to rotate independently of the rack and the ratchet actuating segment. When two or three rounds have entered the mechanism and the first is held between the feed sprockets and the guides inside the case, the tensioning ratchet pawl, engaging the peripheral teeth of the tensioning ratchet, prevents the belt from being withdrawn. The loading is completed by rotating the hexagonal extension of the tensioning ratchet with a spanner or ratchet wrench. Rotation of the ten-

sioning ratchet tends to wind up the driving spring which then reacts through the band brake in the spring case and rotates the shaft and feed sprockets, thus drawing the belt further into the mechanism.

167. As the belt is carried round by the feed sprockets the noses of the bullets are in contact with the bottom of the cam groove formed in the front cover and the rounds are pushed towards the rear. The links are prevented from moving towards the rear by engagement with the teeth of the centre feed sprocket, and when the rounds have been pushed back through a distance equal approximately to the width of a link the links are freed from the rounds. As the rounds are carried past the link ejector they cause the link ejector sprockets to rotate so that the teeth engage the middle portion of each link as it becomes free and push it outwards. The double loops of the link, which are leading, engage the ramp inside the link chute cover with their closed sides and the ramp causes the link to tilt on the teeth of the ejector sprockets so that the single loop is moved clear of the round. Continued movement of the main sprockets carries the round further forward, and the link ejector sprockets rotate so that the single loop of the link engages the ramp inside the link chute cover and the double loops are cleared from the round. The pivotal movement of the link is limited by the engagement of first the double loops and then the single loop with the ramp, so that the single loop cannot re-engage the round and the link falls away through the link chute. After the ejection of the link the spring-loaded plunger in the link ejector bracket, acting on the flats formed on the centre portion of the link ejector shaft, holds the link ejector sprockets in the correct position to be engaged by the next round in the mechanism.

168. Each round, after passing the link ejector, comes into contact with the feed levers and the curved upper surface of the arm on the link ejector bracket. The feed levers are pushed aside, against the action of the return springs and this movement withdraws the last round retainer so that its end is accommodated in the recess in the side of the mouth and does not obstruct the passage of the rounds, which are guided into the mouth by the curved surface of the link ejector bracket arm. After each round passes the feed levers the latter are partly returned by the return springs and when there are two rounds in the mouth the projections on the ends of the levers, bearing on the upper surface of the second round, hold the rounds firmly in the mouth so that the first one can be loaded into the gun without risk of a misfeed. Further manual rotation of the tensioning ratchet, after two rounds have entered the mouth, winds up the driving spring until it reaches its maximum tension, the sprockets and shaft being prevented from rotating by the round which cannot clear the sprockets owing to the rounds in the mouth. The rounds in the mouth are then held in position by the pressure of the third round, which is being urged forward by the sprockets.

169. When the gun is fired, the breech block, as it travels forward, hits the base of the cartridge case of the round which is at the bottom of the mouth and drives the round forward out of the mouth of the belt-feed mechanism into the chamber of the gun, which is thus loaded. As each round leaves the mouth the driving spring, acting through the band brake and the driving spring case, rotates the shaft and the feed sprockets, thereby feeding another round into the mouth.

170. As the gun recoils, the belt feed mechanism, and the magazine carrier on which it is mounted, remain stationary. The bracket and rack-operating lever, mounted on the gun, move backwards, the roller of the rack-operating lever rides up on the inclined face at the front of the magazine carrier so that the rear end of the lever is raised and the rack roller simultaneously rides up the inclined face at the rear of the lever, the combined movement maintaining the driving spring at maximum tension.

171. After the last round but one has been pushed out of the mouth by the breech block the last round begins to move downward. This permits the two feed levers to move to their maximum extent through the action of the rear return springs so that the rear feed lever abuts its stop in the rear cover, and the end of the last round retainer, moving downwards as the front feed lever rotates, rises on the inclined face at the bottom of its recess and projects into the mouth, thus preventing the last round from reaching the loading position. If a fresh belt is fed into the mechanism while the last round of the previous belt is still in the mouth, the first round of the fresh belt becomes the second round in the mouth and the last round of the previous belt can be fired in the normal manner.

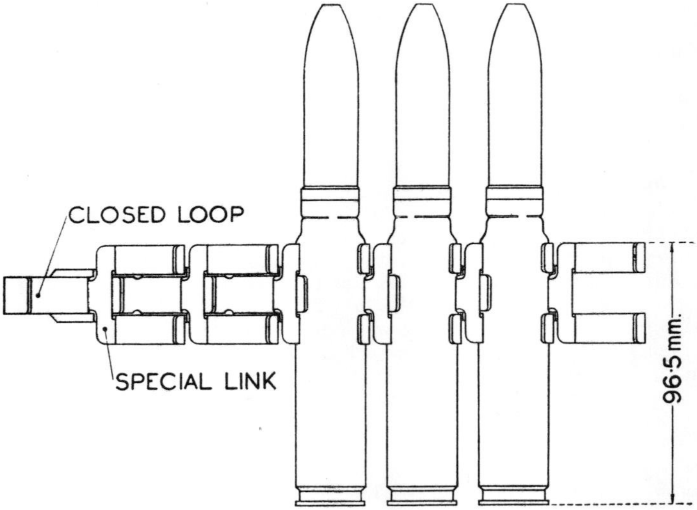

Fig. 25.—Belt, partly assembled, for R.H. feed

Loading and unloading instructions

General instructions

172. Do not apply any oil or grease to the ammunition. Work on a clean surface and keep the links and ammunition free from dirt and dust.

Belt for right-hand feed, fig. 25

173. To make up a belt for right-hand feed, place 27 links in a line along the bench with the double loops to the right and the open side of the loops at the top, the single loop of each link being between the double loops of the next link on the left.

174. Insert a round of ammunition, with the bullet pointing away from the operator, into the loops of each link, including the first (extreme R.H.) link, and push it in until it is about ¼ in. to ⅜ in. short of its final position.

175. Position the rounds accurately in the links. If a positioning machine is not available this can be done by "eye". The distance from the base of the cartridge cases to the front edges of the double loops should be 3·80 in. (96·5 mm.). It will be found that this dimension can be maintained quite accurately once the correct relation of the cartridge case shoulder to the front edge of the link is known.

176. Turn the belt over so that it lies flat on the bench with the backs of the links uppermost. Oil the links lightly with oil, lubricating, anti-freezing, Type A, (Stores Ref. 34A/43 or 46).

177. Test the belt for flexibility by lifting the left-hand loop and drawing it along the back of the belt to the right. In this way the belt will be turned over so that the backs of the links are underneath, and any faulty links will cause the belt to kink instead of folding over smoothly. Any link which does not hinge freely is to be removed and replaced by another, and the test must then be repeated.

178. Repeat the flexibility test detailed in para. 177, starting with the right-hand link and drawing it over to the left. If a stiff link is found, it must be removed and replaced by another and the test must be repeated in both directions.

179. Test the belt for the inclusion of oversize links by suspending it by one end loop and twisting the lowest link until resistance is felt. If the belt breaks due to the single loop of a link jumping off the cartridge case the link is faulty and must be removed and replaced by another, and all the tests must be repeated.

180. Remove the last round at each end of the belt and the corresponding links, leaving a length of belt comprising 25 rounds. These lengths may be joined together by inserting a round in the end links. The round must be pushed in, not snapped into the loops.

181. When the correct length of belt has been assembled, a special end link, with its single loop closed by a bridge piece, must be attached at the trailing (single loop) end. The leading double loop must have a round in it.

Note.—It depends upon the particular aeroplane installation whether it is better to build up the complete length of belt in the filling room or while loading the belt into the aeroplane. It is most important that all the operations detailed in para. 172 to 180 are correctly completed.

Belt for left-hand feed, fig. 26

182. To make up a belt for left-hand feed place 27 links in a line along the bench with the double loops to the left and the open sides of the loops at the top, the single loop of each link being between the double loops of the next link on the right.

183. Insert a round of ammunition, with the bullet pointing away from the operator, into the loops of each link, including the first (extreme L.H.) link, and push it in until it is about $\frac{1}{4}$ in. to $\frac{3}{8}$ in. short of its final position.

184. When the correct length of belt has been assembled, a special end link, with its single loop closed by a bridge piece, must be attached at the trailing (single loop) end. The leading double loop must have a round in it.

185. Complete the operations detailed in para. 175 to 180.

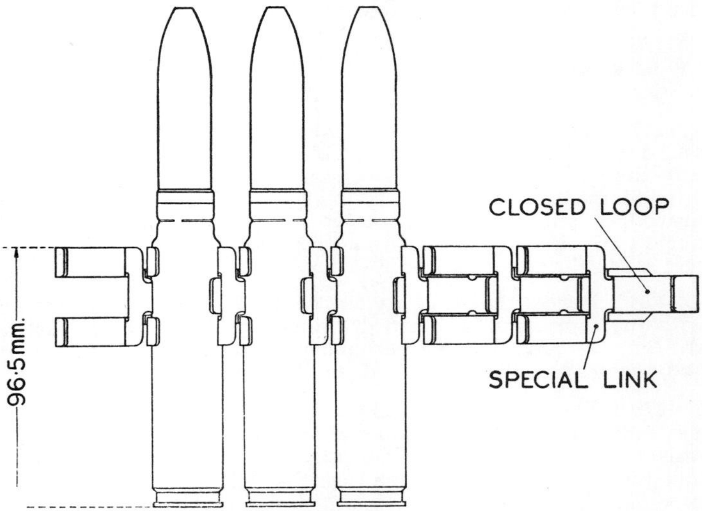

Fig. 26.—Belt, partly assembled, for L.H. feed

To load the gun

186. Ensure that the operations detailed in A.P.1641F, Vol. II, Part 3, Sect. 2, Chap. 4, Group A, have been completed.

187. Cock the gun and set the safety catch of the firing mechanism and the firing button to SAFE.

Note.—These loading instructions are to be applied generally, in conjunction with any specific instructions which may have been issued for a particular aeroplane.

188. Hold the belt feed mechanism above and parallel to the magazine carrier and lower it carefully so that the mouth of the mechanism enters the aperture of the carrier, allowing the shoulder at the rear of the mouth to rest on top of the magazine catch.

189. Press down the front end of the belt feed mechanism and push it forwards so that the two pins at the front of the mouth enter below the hook-shaped projections at the front of the magazine carrier. Raise the lever of the magazine carrier catch, permitting the shoulder at the rear of the belt feed mechanism mouth to fall below the catch, then allow the lever to move down under the action of its springs so that the catch engages the top of the shoulder. Apply lateral and axial pressure to the belt feed mechanism by hand to ascertain that it is properly secured to the magazine carrier.

Note.—When the belt feed mechanism is correctly mounted the rack roller should clear the bottom of the inclined face at the rear of the rack-operating lever by $\frac{1}{32}$ in. to $\frac{1}{16}$ in.

190. Insert the end of the belt into the belt guide of the feed mechanism with the double loops of the links leading and the bullets pointing towards the front. Observe that the open sides of the link loops are facing the mouth of the mechanism as the belt enters the guide. If they do not, the wrong type of belt is being used and it must be withdrawn before the sprockets have engaged the first round, otherwise it will be necessary to dismantle the mechanism.

191. Push the belt as far as possible by hand into the feed mechanism, observing that there is no undue resistance to the entry of the belt and that the rounds are engaged correctly by the sprockets. It should be possible to feed three or four rounds into the mechanism by finger pressure. If the belt does not enter freely, move it in a fore-and-aft direction in the belt guide.

192. Apply a spanner or a ratchet wrench to the hexagonal extension of the tensioning ratchet and rotate the ratchet to draw the belt into the feed mechanism, observing that the ratchet should rotate under a comparatively small load at first and cannot be rotated in the wrong direction owing to the action of the tensioning ratchet pawl.

193. Two links are ejected while the belt is being drawn into the mechanism, and there are then two rounds in the mouth of the mechanism in the correct position for loading the gun.

194. Continue to rotate the tensioning ratchet with the spanner or ratchet wrench until the driving spring is fully tensioned, which will necessitate two or three complete turns of the tensioning ratchet. When the spring is fully tensioned a uniform load will be felt on the spanner or wrench.

195. When the safety catch has been set to FIRE the gun will be ready to fire in the normal manner until the last round only is left in the mouth of the mechanism.

To re-load the gun

196. Cock the gun and set the safety catch of the firing mechanism and the firing button to SAFE.

197. Insert the end of the new belt into the belt guide of the feed mechanism and proceed as detailed in para. 183 to 188,

noting that only one link is ejected down the link chute as the last round of the previous belt remains in the mouth of the mechanism.

198. To join a new belt to a partly expended belt, withdraw the last round of the partly expended belt, thus releasing the special end link, place the single loop of the end link of the partly expended belt between the double loops of the first link of the new belt, ensuring that the open sides of the links of the new belt face the same way as those of the partly expended belt, and insert the free round into the loops of the links at the joint. Ensure that the distance between the base of the cartridge case of the newly inserted round and the front edge of the link is 96·5 mm. and that there is a special link on the end of the new belt.

To unload the gun

199. Ensure that the gun is pointing in a safe direction and that the safety catch of the firing mechanism and the firing button are set to SAFE.

200. Lift the magazine catch lever with one hand, and with the other raise the rear of the belt feed mechanism until the shoulder at the rear of the mouth is clear of the magazine catch. Allow the magazine catch lever to return to its normal position, and, holding the belt feed mechanism with both hands, move it towards the rear, disengage the two pins at the front of the mouth from the hook-shaped projections of the magazine carrier, and lift the mechanism clear of the gun.

201. If the breech of the gun is open, and the chamber is clear, release the breech block.

202. If the breech of the gun is closed, cock the gun and ensure that the chamber is clear, then release the breech block.

To unload the belt feed mechanism

203. Remove the belt feed mechanism from the gun as detailed in para. 200.

204. Withdraw a round from the links near the belt guide to break the belt. Apply a spanner or ratchet wrench to the hexagonal projection of the tensioning ratchet and ensure that the driving spring is fully tensioned.

205. Using the handle of a mallet or other suitable piece of wood, push each round forward in turn out of the mouth, taking care to receive each round as it leaves the mouth so that it is not dropped. Apply further tension to the driving spring during this operation if necessary.

206. To remove the last round from the mouth, open the link chute cover and push the lower end of the front feed lever with a screwdriver or similar implement so that the lever rotates and the last round retainer is withdrawn, permitting the last round to reach the loading position in the bottom of the mouth. Push the round forward and receive it as it leaves the mouth.

6

Browning .50 in Guns

CHAPTER 1

BROWNING ·50 in. GUNS

LIST OF CONTENTS

	Para.
Leading particulars	
Introduction	1
Browning ·50 in. gun, No. 1, Mk. 2	
General description	7
Barrel casing	8
Trunnion block	9
Breech casing	12
Side plates	13
Cartridge stops, link retainer, and retaining pawl	18
Bottom plate	19
Locking piece cam	20
Top plate	21
Back plate with horizontal buffer	22
Breech cover and feed mechanism	23
Feed slide	25
Feed lever	26
Lock frame	27
Oil buffer assembly	29
Accelerator	33
Barrel	34
Barrel extension	35
Breech block locking piece	39
Barrel locking spring	40
Breech block assembly	41
Breech block	42
Transporter and ejector	48
Switch plate	49
Firing pin with extension and spring	50
Sear, sear slide, and sear and firing pin spring retainer	51
Cocking lever	52
Return springs and return springs rod	53
Cocking stud	54
Loading mechanism	55
No. 2 loading mechanism	56
No. 3 loading mechanism	58
Flash eliminator, Mk. 3	61
Action of the gun	64
Recoil action	65
Forward action	70
Safety arrangements	74
Loading and unloading	
Ammunition belt	78
Loading the gun	79
Unloading the gun	88
Instructions for use	94
Browning ·50 in. gun, No. 2, Mk. 2	
General description	98
Back plate	99
Trigger bar	101
Loading mechanism	102
Browning ·50 in. gun, No. 1, Mk. 1	
General description	103
No. 1 loading mechanism	104
Browning ·50 in. gun, No. 2, Mk. 1	
General description	105
Investigation and clearance of stoppages	
General	107
Investigation to be completed by armament personnel	113
If a live round is extracted	121
If an empty case is extracted	126
If nothing is extracted	127

LIST OF ILLUSTRATIONS

Fig.

1. A. Browning ·50 in. gun (M.G.53–2), No. 1, Mk. 2, with No. 2 loading mechanism
   B. Browning ·50 in. gun (M.G.53–2), No. 2, Mk. 2, with No. 3 loading mechanism
2. Browning ·50 in. gun (M.2), No. 1, Mk. 2, with No. 3 loading mechanism
3. Breech casing components and breech cover
4. Back plate with horizontal buffer
5. Breech cover and feed mechanism
6. Lock frame and oil buffer assembly
7. Barrel and barrel extension
8. Breech block, top and bottom views
9. Breech block and components
10. Return springs and return springs rod
11. No. 2 loading mechanism
12. No. 3 loading mechanism with handle
13. Flash eliminator, Mk. 3, and method of locking
14. Recoiling portions in the firing position
15. Recoiling portions in the rear position
16. Back plate Browning ·50 in. gun, No. 2, Mk. 2
17. Browning ·50 in. gun (M.G.53A), No. 1, Mk. 1, with No. 1 standard loading mechanism

BROWNING ·50 in. GUN

Identification of different types of barrel

(i) ·50 in. gun barrels at present in service are known under two main headings, viz.,

(*a*) Barrel, Type D, with sleeve—(two-piece)

(*b*) Barrel, Type E—(one-piece).

(ii) All ·50 in. barrels at present in use in the Service are chromium plated. The chromium plating extends from the muzzle end to within 4·3 to 4·9 in. of the breech end.

(iii) The barrels are identified by a Stores Reference number stamped on each barrel in the position indicated in the figure below.

(iv) Chromium-plated barrels become unserviceable, with existing ammunition, when Plug A/R.169 (Stores Ref. No. 8D/2673) (0·501 in.) enters more than 11·5 in., measured from the extreme end of the plug to the breech face. To facilitate gauging guns *in situ*, a flexible gauging rod (Stores Ref. No. 8D/2750) is provided.

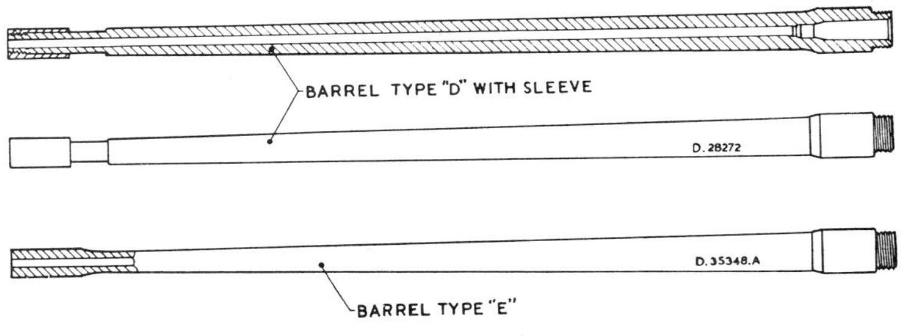

Markings and identification of ·50 in. Browning gun barrels

Leading particulars

Types of ·50 in. Browning guns

American designation	*British number and mark*
M.G.53A (fixed) ...	No. 1, Mk. I (Stores Ref. 107K/89)
M.G.53A (flexible)...	No. 2, Mk. I (Stores Ref. 107K/91)
M.G.53–2 or M2 (fixed) ...	No. 1, Mk. II (Stores Ref. 107K/92)
M.G.53–2 or M2 (flexible) ...	No. 2, Mk. II (Stores Ref. 107K/93)

Means of identification

Mk. I gun
- (i) M.G.53A is stamped on the right side plate below the feedway.
- (ii) The barrel casing is crimped near the muzzle to retain the muzzle bearing.
- (iii) The breech cover catch lever is on the top of the gun.

Mk. II gun
- (i) M.G.53–2, or M2, is stamped on the right side plate below the feedway.
- (ii) The breech cover catch lever is at the side of the gun.
- (iii) The gun has a trunnion block adapter.
- (iv) The back plate catch is provided with a retainer.

Note.—Certain M2 guns of early manufacture have a vertical buffer in the back plate, in other respects the M2 gun is the same as the M.G.53–2 gun.

Length of gun, overall ...	4 ft. 8 in.
Length of barrel ...	3 ft. 6 in.
Weight of gun ...	Mk. I, 55 lb. to 58 lb. Mk. II, 64 lb. to 68 lb.
Weight of barrel ...	Mk. I, 7 lb. or 9½ lb. Mk. II, 9½ lb. or 13½ lb.
Rifling	
(i) Type ...	Uniform right-hand
(ii) Twist ...	1 turn in 15 in.
(iii) No. of grooves ...	8
Method of feed ...	Belt
Ammunition ...	·50 in. rimless
Weight of belt, per 100 rounds ...	30 lb. 4 oz., approx.
Muzzle velocity ...	2,530 f.s., approx.
Speed of fire ...	800 rounds per min., approx.
Space above the gun required to remove the back plate ...	5·2 in.
Space to rear of gun required to remove the lock frame and breech block ...	9·1 in.
Space to rear of gun required to remove the barrel extension and barrel, having removed the lock frame ...	47 in.
Space above the gun required to open the breech cover	13·5 in.

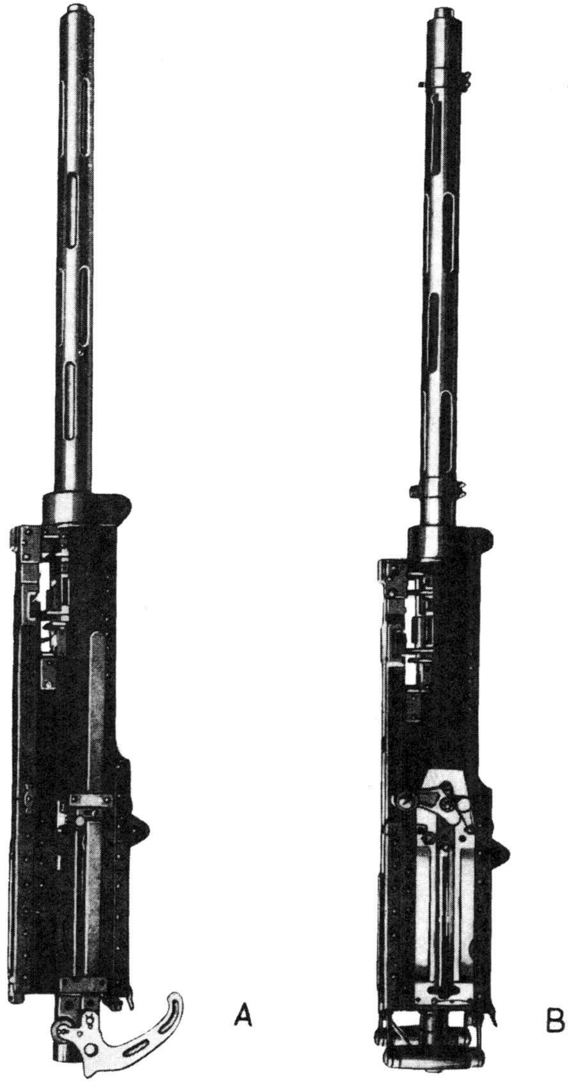

Fig. 1.—A. Browning ·50 in. gun (M.G.53–2), No. 1, Mk. II, with No. 2 loading mechanism

B. Browning ·50 in. gun (M.G.53–2), No. 2, Mk. II, with No. 3 loading mechanism

Introduction

1. The ·50 in. Browning gun is designed for use in fixed or free positions in an aeroplane, or it can be used for ground firing. The No. 1, Mk. 1, or the No. 1, Mk. 2, gun is mounted in a fixed position, and the No. 2, Mk. 1, or No. 2, Mk. 2, gun is mounted in a free position. The No. 2, Mk. 1 and 2 guns are also used for ground firing. The Mk. 2 gun differs slightly from the Mk. 1 gun in construction and, in each mark of gun, the No. 2 gun differs from the No. 1 gun in being equipped with a different back plate, and different type of loading mechanism, (*fig. 1.*)

2. When mounted in fixed positions to fire through the plane of rotation of the propeller, the firing of the gun is controlled by a synchronizing mechanism which operates a trigger motor on the gun. When mounted in a fixed position in the wing, the gun is fired by an electric firing unit fitted in place of the trigger motor. In both instances firing is effected by closing a firing switch on the control column. The trigger motor or the electric firing unit can be mounted on either side of the gun, to suit the installation. The synchronizing mechanism and electric firing unit are described in A.P.1641N, Vol. 1. When mounted in free positions or when used for ground firing, the gun is equipped with a back plate having a double hand grip and is fired direct by mechanical means, a trigger mounted in the back plate operating a trigger bar in the breech casing which engages the sear. The gun is not provided with a rear sear and stops firing with a live round in the chamber.

3. The gun is cooled by air flow over the barrel and when used for ground firing not more than 50 rounds should be fired without allowing the gun to cool.

4. The ·50 in. Browning gun is automatic in action in that, so long as the firing control mechanism is operated, the gun should continue to fire until all the available ammunition is expended. Ammunition is fed to the gun by a belt which may be of any length, depending upon the stowage space available and the power of the gun to feed in the belt. Provision is made for left-hand or right-hand feed, to suit the installation, and the change from one feed to the other can be readily effected by repositioning certain parts of the feed mechanism.

5. The speed of fire of the gun, which is approximately 800 rounds per min., can be adjusted by regulating the flow of oil in an oil buffer.

6. The gun depends upon the force of recoil for its automatic action. As each round is fired the barrel and barrel extension recoil a short distance and the movement unlocks the breech and throws the breech block to the rear, compressing the breech block return spring. The rate of recoil of the barrel, and therefore the rate of fire of the gun, is controlled by an oil buffer and, in recoiling, the barrel compresses the barrel return spring. As the breech block moves to the rear it extracts the empty cartridge case from the chamber, and at the same time withdraws a live round from the belt, the empty cartridge case being ejected at the end of the movement, through the bottom of the breech casing. The breech block rebounds off a buffer in the back plate and is returned by its spring to load the round into the chamber. When the breech block reaches the closed position, the barrel, barrel extension, and breech block are returned to the firing position, the movement being assisted by the impulse of the barrel return spring, and the breech block is positively locked to the barrel extension. The feed mechanism is operated by the reciprocating movement of the breech block, and if the firing control mechanism is operated this cycle will be repeated.

BROWNING ·50 in. GUN, No. 1, Mk. 2

GENERAL DESCRIPTION

7. The gun may be divided naturally into three parts for purposes of description:—

(i) The non-recoiling portions, which consist of:—
- (*a*) The barrel casing,
- (*b*) The trunnion block,
- (*c*) The breech casing with incorporated fittings, and back plate, and breech cover with feed mechanism,
- (*d*) The lock frame which houses the oil buffer, barrel return spring, and accelerator.

(ii) The recoiling portions, which consist of:—
- (*a*) The barrel and barrel extension, with breech block locking piece, and barrel locking spring,
- (*b*) The breech block with transporter, switch plate, and firing mechanism,
- (*c*) The breech block return spring and cocking stud.

(iii) The loading mechanism.

For details of functioning, safety arrangements, loading and unloading and operations in the air, see para. 61 to 94.

Barrel casing, fig. 1 and 2

8. The barrel casing is a drawn steel tube which has a number of slots cut in it to allow air access to the barrel for cooling purposes. At the rear of the barrel casing is a support or bearing which screws into the front of the trunnion block and holds the casing in position. Screwed into the front of the barrel casing, and secured by set-screws, is a muzzle bearing which supports the muzzle of the gun.

Note.—The barrel casing of the M2 gun has a number of circular holes instead of slots.

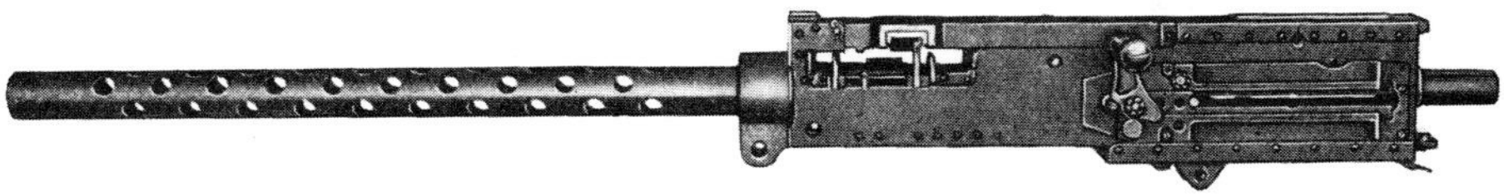

Fig. 2.—Browning ·50 in. gun (M2) No. 1, Mk. 2, with No. 3 loading mechanism

Trunnion block, fig. 3

9. The trunnion block is flat on each side for the attachment of the breach casing side plates, which are riveted in position. At the front of the trunnion block is a cylindrical boss which is threaded internally for the rear support of the barrel casing, and externally for an adapter. Above and to the rear of the cylindrical boss two wings form a gap to receive the hinge of the breech cover. Between the wings is positioned a spring-loaded breech cover detent and at the top is fitted a trunnion block cover which is secured by two pins.

10. The trunnion block is bored longitudinally to receive and form a rear bearing for the barrel, and on the top at the rear it is machined to form the feedway, slots being cut for the front and rear cartridge stops, link retainers, and retaining pawl.

11. The trunnion block adapter screws on to the front of the trunnion block, being positioned by a shim and prevented from unscrewing by a spring-loaded plunger which is secured by a split pin. A hole is bored through the underside of the adapter for the gun front mounting bolt.

Breech casing, fig. 3

12. The breech casing is a rectangular box made up of two side plates and a top and bottom plate. The top and bottom plates are shorter than the side plates, and at the front, the breech casing is open at the top and at the bottom. The top is closed by the breech cover whilst the bottom is permanently open and forms the ejection orifice. The rear of the breech casing is closed by the back plate which is attached by guides and grooves and is retained by a catch. All the working parts of the gun with the exception of the barrel, are housed in the breech casing. A stop barrel locking spring is riveted on the inner side of the right-hand side plate to prevent the barrel locking spring from disengaging the recesses in the barrel when the flash eliminator is unscrewed.

Side plates

13. The side plates are riveted at the front to the trunnion block and are cut away to conform to the feedway. Immediately below the feedway, and riveted to each side plate, is a retaining pawl bracket which is machined to accommodate the cartridge stops, link retainer, retaining pawl, and lugs for the ammunition box neck and link chute. These components are retained in position by spring-loaded pins and their position on the right or left of the gun depends upon the direction of feed.

14. In each side plate is cut a slot to accommodate the cocking stud which moves backwards and forward with the breech block, the middle of each slot being cut away to allow the cocking stud to be assembled and dismantled. Above the front end of each cocking stud slot are two slots for the assembly of the trigger motor or electric firing unit. The trigger motor or electric firing unit must always be fitted on one side of the gun and the cocking stud and loading mechanism on the other, but they can be reversed to suit the installation. Holes are drilled in each side plate at the front and rear of the cocking stud slot for the loading mechanism guide brackets. There are three holes in each side plate for securing a gun heater to the gun.

15. A hole is cut in the rear of the right side plate at the top to accommodate a projection on the return spring rod head. Below this hole is a second hole through which a punch can be inserted to disengage the lock frame retainer when dismantling the gun. A hole for the trigger bar pin is drilled in the left side plate near the top. At the rear of each side plate is a vertical groove to accommodate the corresponding guide on the back plate.

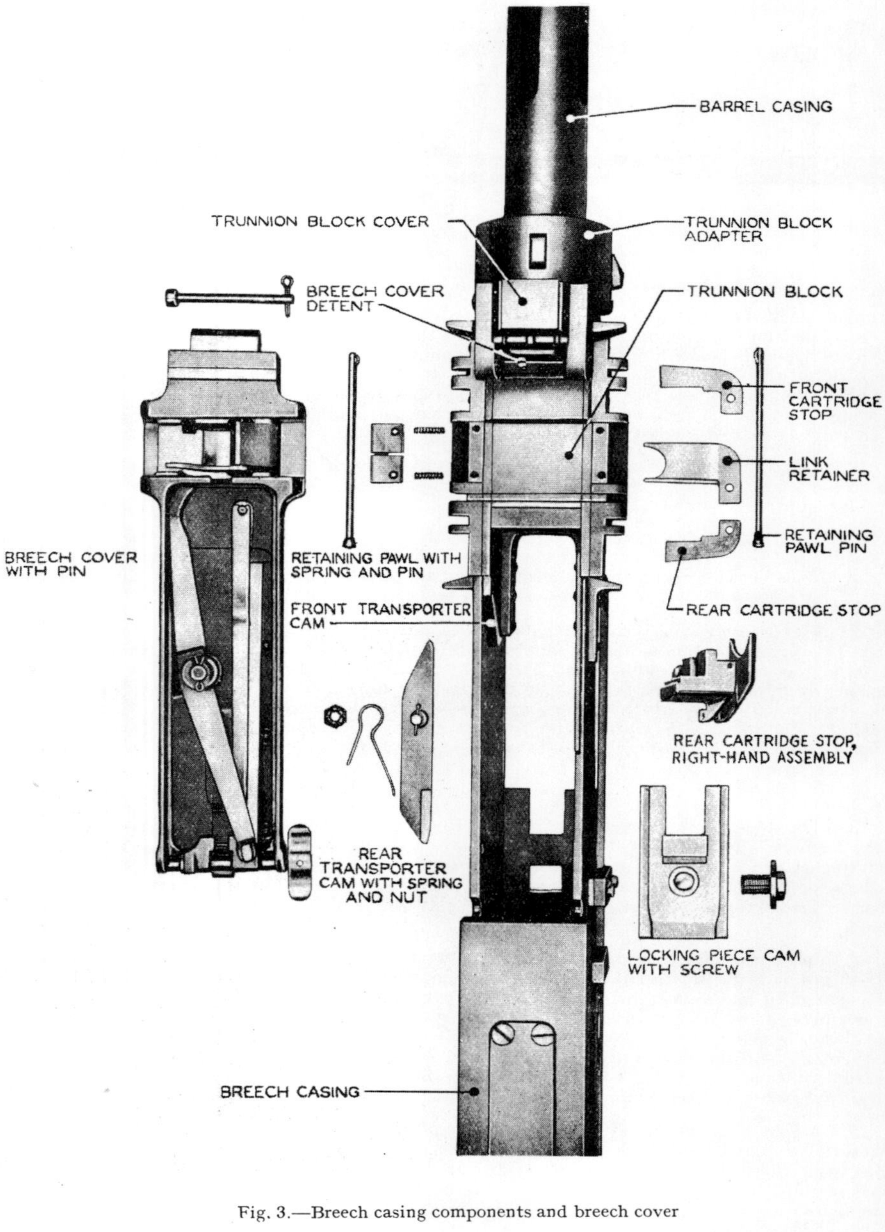

Fig. 3.—Breech casing components and breech cover

16. A short longitudinal slot is cut inside each side plate at the rear to accommodate projections on the lock frame, and above the slot in the right side plate is a longer slot for the lock frame retainer, Riveted to the upper rear end of the right side plate is a rib which prevents the breech block from moving upward as it travels backwards and forwards.

17. A front transporter cam is riveted to the top of the left side plate to the rear of the trunnion block, and to the rear of this cam is a rear transporter cam which is spring-loaded and pivots on a stud passing through the side plate, the stud being secured by a castellated nut and split pin. These cams help to govern the movement of the transporter as the breech block moves backwards and forwards.

Cartridge stops, link retainer, and retaining pawl, fig. 3

18. The cartridge stops are L-shaped and project into the feedway and limit the travel of the belt. The link retainer is also L-shaped and is positioned between the cartridge stops. Its inner end is shaped to conform to the contour of the cartridge case, which it engages just behind the link. As the round is withdrawn from the belt the link retainer prevents the movement of the link to the rear. In figure 3 are shown the normal front and rear cartridge stops and link retainer, and also a rear R.H. cartridge stop assembly. When the gun is assembled for L.H. feed it is preferable to fit a rear R.H. cartridge stop assembly instead of the normal rear cartridge stop and link retainer. The retainer pawl assembly, which is always fitted in the feedway on the opposite side to the cartridge stops, consists of a right-hand and a left-hand pawl connected together by a sleeve. Each pawl is spring-loaded.

Bottom plate

19. The bottom plate is riveted to the side plates and closes the underside of the breech casing at the rear. It accommodates the locking piece cam at its forward end, the cam being secured by a screw. On the bottom plate two projections are formed and drilled for the bolt of the gun rear mounting.

Locking piece cam, fig. 3

20. The locking piece cam is a rectangular block having guides on each side to mate with corresponding grooves in the barrel extension and prevent this from lifting during its backward and forward movement. On the top of the locking piece cam at the front is a projection, the rear face of which is bevelled to engage and lift the locking piece as the recoiling portions move forward. In front of the bevel is a flat which retains the locking piece in the locked position during firing.

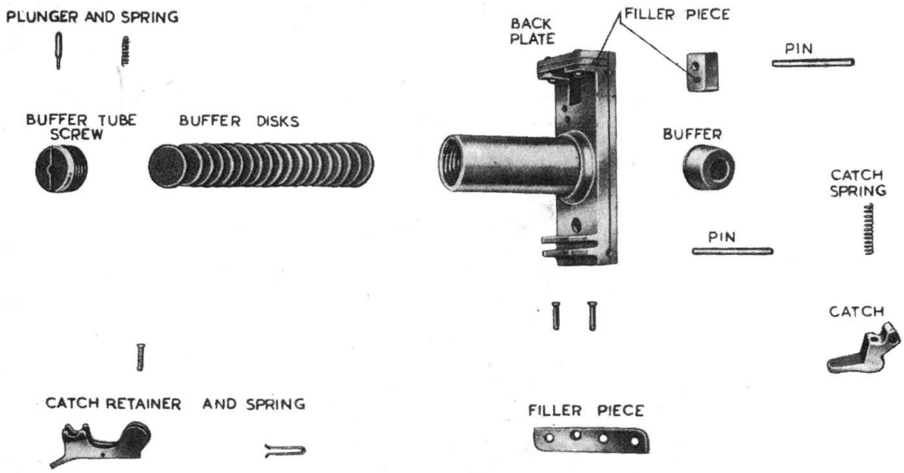

Fig. 4.—Back plate with horizontal buffer

Top plate

21. The top plate is riveted to the side plates towards the rear and carries on its underside a bracket which actuates the cocking lever as the breech block slides past. At the rear of the top plate is a stud and at the front of the top plate bracket is a second stud. These studs form rear and front stops for the trigger bar, which is mounted in the top plate bracket when the gun is used as a No. 2 gun.

Back plate with horizontal buffer, fig. 4

22. The back plate is attached to the rear of the breech casing by guides and grooves and is retained by a spring-loaded catch which engages the bottom plate. The catch is prevented from becoming disengaged by a spring-loaded retainer. The back plate forms a rear abutment for the breech block return spring rod and the lock frame and takes the shock of recoil which is transferred to the trunnion lock by the side plates. Projecting to the rear of the back plate is a tube which accommodates a buffer at its front end. Behind the buffer are 22 fibre discs which are retained by a screw secured by a spring-loaded plunger. At the top and bottom of the back plate, filler pieces are inserted. These filler pieces are removed when the back plate is equipped with hand grip and trigger for use as a No. 2 gun.

Breech cover and feed mechanism, fig. 3 and 5

23. The breech cover is hinged at its forward end to the top of the trunnion block and, in addition to closing the forward part of the breech casing, carries the feed mechanism. When closed, the breech cover is retained by a spring-loaded catch which engages the front of the top plate. On the outer end of the catch pin is a lever which enables the catch to be disengaged and the cover opened. Transverse slots are cut at the front end of the cover to engage a spring-loaded detent which holds the cover in the open position. At the front, the cover is shaped to accommodate the feed slide, grooves being formed in the cover to mate with guides on the slide.

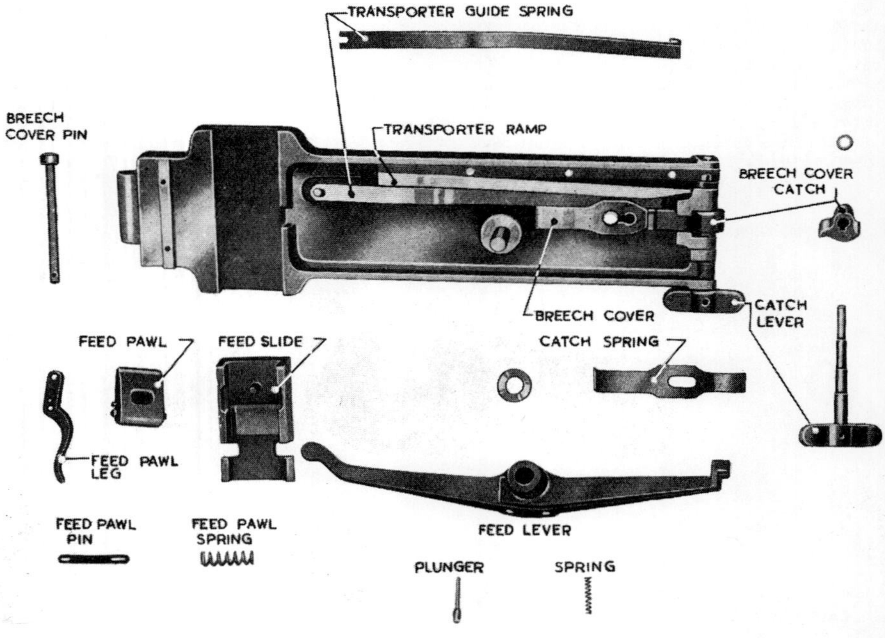

Fig. 5.—Breech cover and feed mechanism

24. A transporter ramp is riveted to the cover near the left-hand side. The ramp forces the transporter downwards as the breech block travels to the rear. By the side of the ramp is mounted a transporter guide spring, the front end of the spring engaging a stud in the breech casing, and the rear end a slot in the transporter ramp. The spring ensures engagement of the transporter with the round in the feedway when the breech block moves forward. A stud for the feed lever is also riveted to the inside of the cover.

Feed slide

25. The feed slide, which moves laterally in the breech cover, carries a spring-loaded feed pawl and has a slot cut in it for engagement by the end of the feed lever. Attached to the feed

pawl is a leg, which prevents the feed pawl from engaging a round if there are already two rounds in the feedway, thus avoiding double feed.

Feed lever

26. The feed lever is held in position on its stud in the breech cover by a washer and split pin. Two holes are drilled in the feed lever near its axis and into one of these holes, depending upon the direction of feed, is assembled a plunger and spring. The plunger retains the feed lever in the correct position for engaging the appropriate groove in the breech block providing this is closed when the cover is lowered. At one end of the feed lever is a stud which engages the groove in the breech block, the other end of the feed lever being shaped to engage the feed slide.

Lock frame, fig. 6

27. The lock frame which accommodates the oil buffer and accelerator, is held in position in the rear of the breech casing by a spring retainer. When the back plate is in position it forms the rear abutment for the oil buffer and prevents any shock from being taken by the lock-frame spring retainer. On each side of the lock frame is a guide which mates with the slot in the side of the breech casing. The lock frame is bored from the rear to form a housing for the buffer cylinder, a flange being formed about half-way along it to provide a limit stop for the cylinder. Accommodated in undercuts in the under side of the lock frame is a detent spring which has a stud at its rear end to engage serrations in the buffer cylinder and prevent this from turning. During the forward and backward movement of the breech block the front end of the detent spring engages the accelerator and prevents this from moving.

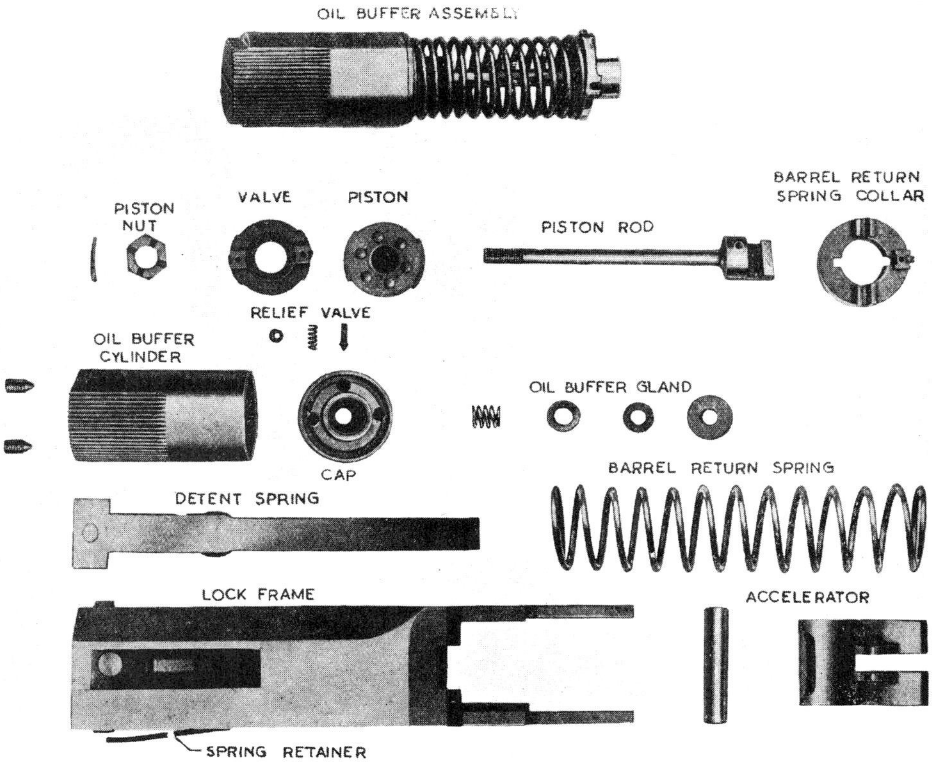

Fig. 6. -Lock frame and oil buffer assembly

28. On each side of the lock frame, at the front, is a projection in which is accommodated a locking piece depressor, the depressors being secured by rivets. At the front end of the depressor is a bevel which engages the locking piece pin as the gun recoils, and lowers the pin and locking piece to disengage the latter from the breech block.

Oil buffer assembly, fig. 6

29. The oil buffer assembly consists of a cylinder with cap and gland, piston rod with piston and valve, barrel return spring, and barrel return spring collar.

30. Filling and emptying holes are situated in the rear of the buffer cylinder; these holes are closed by screwed plugs and copper washers. The front of the cylinder is closed by a screwed cap accommodating a gland through which the piston rod passes. The gland consists of a four-coil spring, bush, and packing held in position by a nut. A spring-loaded relief valve is also accommodated in the cylinder cap.

31. The piston is screwed on to the inner end of the piston rod and is secured by a pin. The valve is loosely mounted on a boss on the piston and is retained by a slotted nut which is secured by the piston pin. The piston and valve exactly fit the bore of the cylinder but their edges are cut away to form ports for the flow of oil during recoil. On the valve are two studs which engage grooves in the cylinder, so that when the cylinder is rotated the valve will rotate on the piston and vary the size of the ports, thus controlling the flow of oil. Six holes are drilled through the piston. The holes are closed by the valve during the recoil of the gun but are opened to the free passage of oil during the return movement.

32. The outer end of the piston rod is enlarged in diameter and is shaped and slotted to engage a projection on the barrel extension. Passing through the enlarged portion of the piston rod is a pin which holds the barrel return spring collar in position, the collar being prevented from turning by a stud which engages a slot in the lock frame. The barrel return spring surrounds the piston rod and bears between the collar and the buffer cylinder cap.

Accelerator, fig. 6

33. The accelerator is mounted on a pin at the forward end of the lock frame and has two curved horns, each with a ramp on the rear concave face. The horns operate on each side of the barrel extension shank and, when the accelerator is rotated to the rear, the ramps engage the shank and lock the barrel extension to the lock frame. In the underside of the accelerator is a small recess in which the buffer cylinder detent spring engages when the accelerator is in its rear position.

Barrel, fig. 7

34. The barrel is bored and rifled and at the rear is a chamber which conforms to the contour of the ·50 in. rimless cartridge case. Between the chamber and commencement of the rifling is formed a lead. Externally at the breech end and muzzle, the barrel is cylindrical in shape and, to the muzzle is attached a barrel sleeve. The breech end of the barrel is threaded externally to screw into the barrel extension and in front of the threaded portion are a number of short longitudinal V-notches for engagement with the barrel locking spring. The thread enables the cartridge head space to be adjusted, the barrel being locked after adjustment by the locking spring.

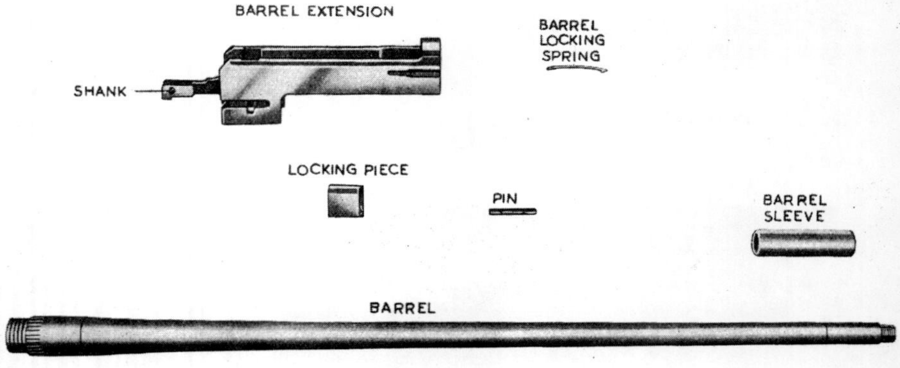

Fig. 7.—Barrel and barrel extension

Barrel extension, fig. 7

35. The barrel extension consists of two side members which are integral with a front portion, and a downward projecting lug at the rear. The front portion is threaded internally for the

attachment of the barrel, and the lug has a vertical slot cut in it to accommodate the breech block locking piece; an elongated transverse hole in the lug allows the locking piece pin to move vertically. Pinned to the rear of the lug is a shank which engages the oil buffer piston.

36. The under side of the lug is recessed to clear the locking piece cam, and grooves are formed in the lug to engage the guides on the cam. The guides and grooves prevent any tendency of the barrel extension to lift when the locking piece rides up the cam at the end of the forward movement of the recoiling portions.

37. In the right-hand side of the front of the barrel extension is a slot with undercuts, which accommodates the barrel locking spring, (see para. 40) and in each side at the rear, clearance slots are cut to accommodate the locking piece depressors of the lock frame.

38. Running the length of the side members, in the inside, are grooves in which guides on the breech block slide as the breech block travels backwards and forwards.

Breech block locking piece, fig. 7

39. The locking piece, which locks the breech block to the barrel extension in the firing position, is a rectangular block having a transverse hole drilled through it for the locking piece pin. The top front face is bevelled to provide a lead into the locking recess in the breech block in the event of the breech block not being fully home when the locking piece engages the locking-piece cam. The locking piece is also bevelled at the bottom where it engages the locking-piece cam.

Barrel locking spring, fig. 7

40. The barrel locking spring is a flat steel spring having a projection at its forward end to engage the notches at the breech end of the barrel.

Breech block assembly

41. The breech block assembly consists of the breech block, transporter with ejector, switch plate, firing mechanism, breech block return springs and return springs rod, and cocking stud. The firing mechanism consists of a firing pin extension, firing pin spring, Mk. 1, sear, sear slide, sear and firing pin spring retainer, and cocking lever. The functions of the moving breech block assembly are loading, firing, extracting, and ejecting the empty cartridge case and providing the energy to operate the feed mechanism.

Breech block, fig. 8 and 9

42. The breech block is supported in the barrel extension by corresponding guides and grooves and has a longitudinal movement in line with the axis of the bore. Its rearward movement is limited by the back plate buffer and its forward movement by the breech face of the barrel.

43. Cut in the underside of the breech block, towards the rear, is a recess in which the locking piece engages when the breech is closed and locks the breech block to the barrel extension. Behind the recess, clearance is cut for the horns of the accelerator, the rear face of the clearance projecting downwards to form a lug for engagement with the accelerator. Through the lug a hole is drilled for the cocking stud.

44. Two interrupted cam grooves, for the feed lever, one for right-hand feed and the other for left-hand feed, are cut in the top of the breech block. The appropriate groove to suit the direction of feed is made continuous by the switch plate. The top of the block is recessed to accommodate the switch plate and a hole is drilled for the switch plate plunger. Towards the rear of the breech block a slot is cut for the cocking lever and an undercut for the plate of the sear and firing pin spring retainer, a vertical hole being drilled through the breech block for the pin of the retainer. A transverse hole is drilled through the breech block for the cocking lever pin.

45. On each side of the breech block there is a guide to mate with the groove in the side member of the barrel extension. A hole is drilled in the left-hand side of the breech block for the transporter axis, the hole breaking through into the axis hole for the switch plate. An undercut, for the flange of the transporter, is formed in the breech block near the transporter axis hole. Towards the front of the breech block is a stop which limits the downward movement of the transporter.

46. The front face of the breech block is prepared with undercut guides to form a cartridge way, the tops of the guides being chamfered to facilitate engagement by the base of the cartridge case.

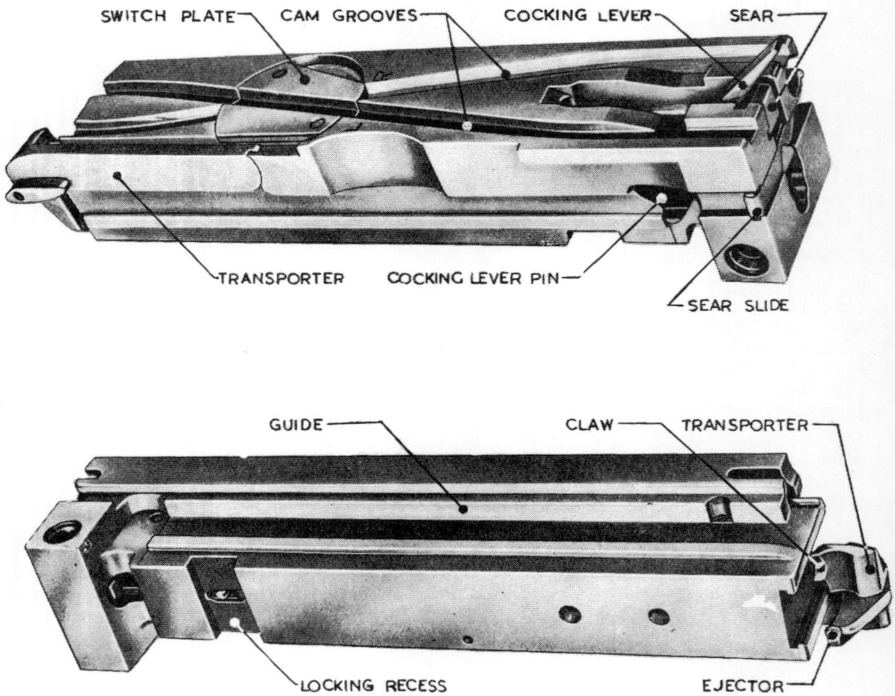

Fig. 8.—Breech block, top and bottom views

47. The rear of the breech block is recessed and grooved vertically for the sear and horizontally for the sear slide. Two longitudinal holes are drilled for the firing pin and return spring, respectively.

Transporter and ejector, fig. 9

48. The transporter has an integral axis which pivots in the side of the breech block, a radial flange near the axis engaging the undercut in the breech block and holding the transporter in position. At the other end of the transporter is a projection with a claw and mounted in the projection is a spring-loaded ejector. On the outside of the transporter is a guide which contacts the front and rear transporter cams of the breech casing as the breech block travels backwards and forwards.

Switch plate, fig. 9

49. The switch plate is circular in shape, being cut away on each side to conform to the contour of the breech block, and has an integral axis. A groove is cut in the top of the plate to complete the cam grooves in the breech block, two holes being drilled in the plate for the head of the plunger which retains the switch plate in the correct position for the required direction of feed. A circumferential groove is cut in the axis, the groove being engaged by the transporter axis, which thus retains the switch plate in the breech block.

Firing pin, with extension and spring, fig. 9

50. The firing pin is cylindrical in shape and decreases in diameter towards the front to form the pin. At the rear, the firing pin is enlarged and has a circumferential groove cut in it for engagement with the firing pin extension. The firing pin extension is drilled to accommodate the firing pin spring, which is held in position by a pin. The rear of the firing pin spring bears against the pin

of the sear retainer which passes through a slot in the rear of the firing pin extension and a hole in the breech block. The rear end of the firing pin extension is shaped on its under side to form a bent for engagement with the sear. The firing pin and firing pin extension are fluted to prevent any cushioning effect on the two components when they are released by the firing gear.

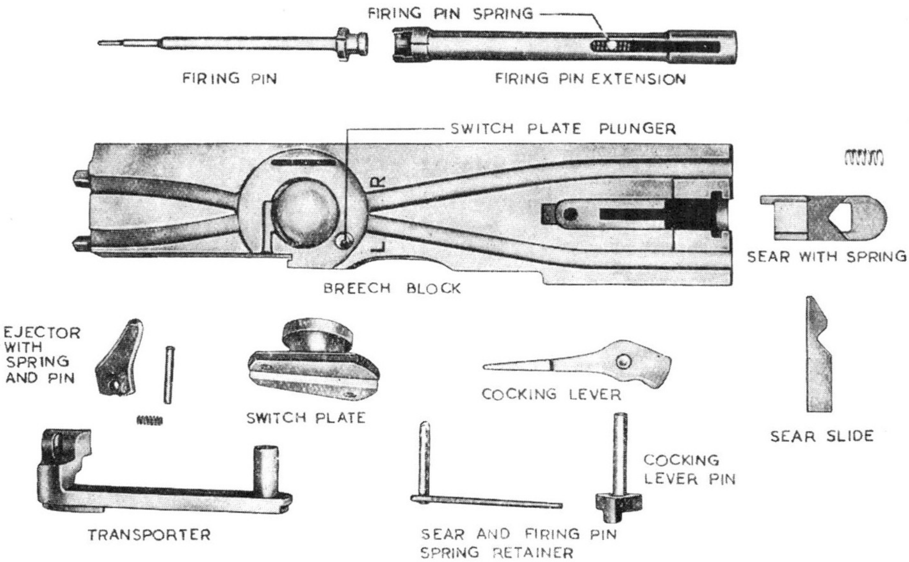

Fig. 9.—Breech block and components

Sear, sear slide, and sear pin and firing pin spring retainer, fig. 9

51. The sear, which is accommodated in the vertically-grooved recess in the rear of the breech block, is spring-loaded and has a bent on its front face. The spring bears between the underside of the sear and the bottom of the recess. The sear slide is accommodated in the horizontally-grooved recess and has a notch in its lower edge which engages a cam on the back of the sear. The sear slide can be inserted from either side of the breech block to suit the position of the trigger motor or electric firing unit. The sear is held in position by a retainer which consists of a plate and pin. The plate engages an undercut in the breech block, in which position it is held by the cocking lever, and the pin passes through a hole in the breech block and a slot in the firing pin to form a rear abutment for the firing pin spring.

Cocking lever, fig. 9

52. The cocking lever pivots on its pin in the breech block, a cam on its lower end operating in the slot in the firing pin extension. The upper end of the cocking lever engages the top plate bracket of the breech casing.

Fig. 10.—Return springs and return springs rod

Return springs and return springs rod, fig. 10

53. The return springs consists of an outer and an inner spring mounted on a rod. The rod gives the springs support and enables initial compression to be applied. On the rear of the rod is

a head having a projection to engage the hole in the right side plate of the breech casing. At the front end of the rod a collar, which is retained by a pin, holds the springs under initial compression. The return springs rod is free to pass through the hole in the breech block when the breech block recoils but the springs are seated near the front of the breech block by the collar and are further compressed by the recoiling breech block.

Cocking stud

54. The cocking stud can be assembled on either side of the breech block to suit the position of the loading mechanism, its function being to transfer the movement of the loading mechanism to the breech block. A groove is turned at one end of the cocking stud to retain the loading implement in engagement with the stud when the gun is being cocked.

Loading mechanism

55. Two types of loading mechanism are provided for the No. 1, Mk. 2, ·50 in. Browning gun, the No. 2 loading mechanism which is used when the gun is mounted in the cockpit within the reach of the pilot and the No. 3 loading mechanism which is used when the gun is mounted outside the reach of the pilot or when the gun is mounted in the wing. The loading mechanism can be positioned on either side of the gun to suit the installation, but must always be fitted on the opposite side tc the trigger motor or electric firing unit.

No. 2 loading mechanism, fig. 11

56. The No. 2 loading mechanism consists of an operating bar which slides in front and rear guide brackets. The guide brackets are fixed to the side plates of the breech casing by screws which are secured by locking wires. Near the middle of the operating bar is a hook for engagement with the cocking stud, and near the rear end is a stop to limit the forward movement of the mechanism. In the underside of the operating bar are three notches, one of which will engage the rear guide bracket and retain the loading mechanism in the forward or rear position, two springs being positioned in the top of the bar to prevent disengagement.

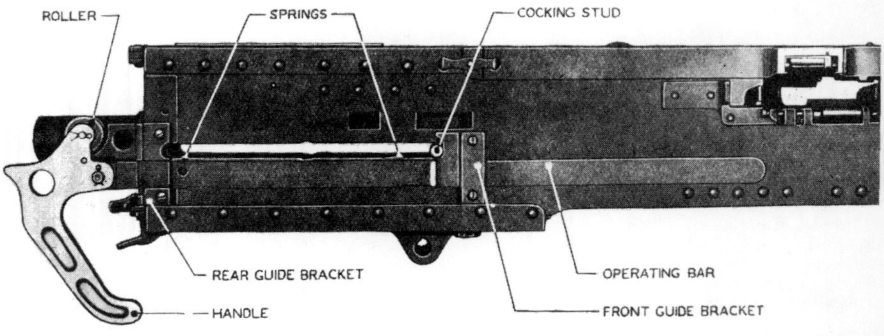

Fig. 11.—No. 2 loading mechanism

57. Attached to the rear of the operating bar by a hinge pin is a handle in which is mounted a roller. A spring-loaded plunger is accommodated in a hole in the handle to retain the handle in its normal position. When the handle is lifted to open the breech, the roller engages a cam on the rear guide bracket and provides the necessary leverage to withdraw the barrel and barrel extension, and unlock the breech block.

No. 3 loading mechanism, fig. 12

58. The No. 3 loading mechanism consists of a slide bracket which is fixed to the side plate of the breech casing by two screws at the rear and one screw and a bolt at the front. The front screw is longer than the rear screws and the bolt engages the trigger motor slot. The slide bracket is slotted for almost its whole length and forms a guide for the operating slide. At the front end of the slide bracket is a screwed stud which provides a fulcrum for the operating lever. In the front of the side bracket is accommodated a spring-loaded plunger which retains the operating slide in the forward position.

59. The operating slide is flanged to engage the bracket and has a notch at the rear to engage the cocking stud and a notch in the top to engage the retaining plunger. Screwed into the slide is a stud, which forms a pivot for the operating lever, and also a stop for the operating lever.

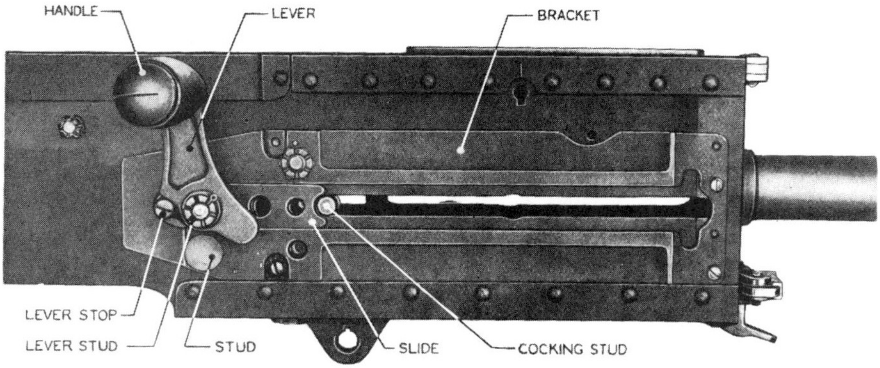

Fig. 12.—No. 3 loading mechanism with handle. (The handle is removed when used with a No. 1, Mk. 2 gun)

60. The operating lever is held in position by a slotted nut with washer and split pin. A torsion spring, one end of which is anchored to the stop and the other end to the lever, retains the lever in its normal position. Two torsion springs are provided, one for use when the loading mechanism is mounted on the left of the gun and the other for use when the loading mechanism is mounted on the right of the gun. A hole is drilled in the end of the lever and threaded for the attachment of the handle, which is mounted on a tube and secured by a bolt. The handle is for use when the loading mechanism is fitted to the No. 2, Mk. 2, ·50 in. Browning gun. A remote control cable is attached to the end of the lever when the loading mechanism is fitted to the No. 1, Mk. 2 gun.

Flash eliminator, Mk. 3, fig. 13

61. An eliminator, flash, Mk. 3 (Stores Ref. 107K/549) is provided for fitting to the muzzle of the gun when used in certain installations, particularly in aircraft turrets. It can only be used on guns fitted with the Type "D" barrel; the rear end of the flash eliminator serves in place of the barrel sleeve.

Note.—The Eliminator, flash, Mk. 3, supersedes the Eliminator, flash, Mk. 2 (Stores Ref. 107K/489).

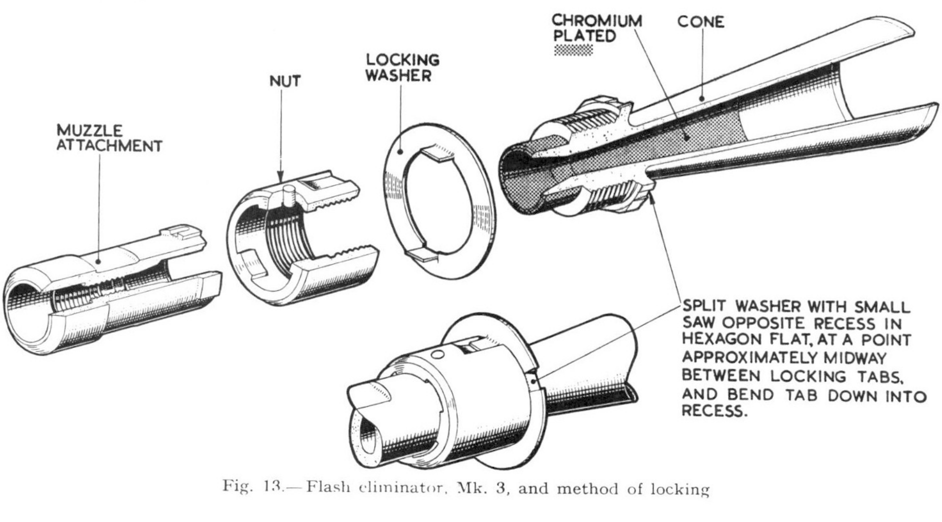

Fig. 13.—Flash eliminator, Mk. 3, and method of locking

62. The components of the flash eliminator comprise a cone, muzzle attachment, nut, and locking washer. The cone incorporates an integral choke chamber at the rear end, the inside and outside of which, together with the rear end of the inside of the cone, are chromium plated to resist wear and erosion. The outside diameter of the choke chamber forms a spigot which seats in the front end of the muzzle attachment; the cone and muzzle attachment are held together with the nut, which is screwed up as tightly as possible, and securely locked by the locking washer, as shown in fig. 13.

63. The rear end of the muzzle attachment is screwed as tightly as possible on to the threaded end of the Type "D" barrel in place of the barrel sleeve, the outside diameter of the attachment locating in the Barrel-housing, front, Mk. II (Stores Ref. 107K/153).

ACTION OF THE GUN

64. When the gun is fired at its maximum rate there is approximately 1/13 sec. between the firing of each round and the various movements take place almost simultaneously.

Recoil action, fig. 14

65. When the gun is loaded and the sear slide is pushed inwards by the trigger motor or electric firing unit, the notch in the sear slide engages the cam on the sear and forces the sear downwards compressing the sear spring and releasing the firing pin. The firing pin flies forward and strikes the cap to fire the cartridge.

66. Upon the discharge of the cartridge, the barrel recoils carrying with it the barrel extension and breech block, which remains positively locked during the first part of the recoil. After the parts have recoiled about $\frac{3}{4}$ in., the locking piece pin is engaged by the bevels at the front of the depressors, the locking piece having reached the top of the incline of the locking piece cam. The

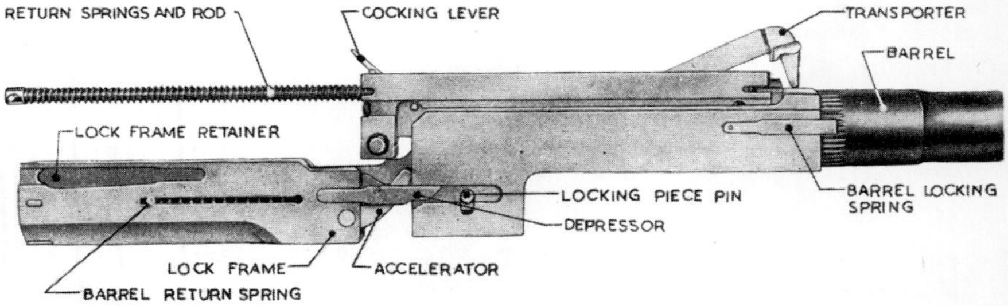

Fig. 14.—Recoiling portions in the firing position

continued movement of the recoiling portions causes the depressors to force the locking piece downwards, thus disengaging it from the breech block and unlocking the breech block from the barrel extension. At the same time, the barrel extension strikes against the horns of the accelerator, causing the accelerator to rotate backwards and engage the lug at the rear of the breech block. The accelerator separates the breech block from the barrel extension and throws the breech block to the rear, compressing the breech block return spring. As the accelerator rotates it engages the barrel extension shank and locks the barrel extension to the lock frame, the accelerator being retained by the oil buffer cylinder detent spring.

67. The recoil of the barrel and barrel extension causes the barrel extension shank to force the oil buffer piston to the rear, compressing the barrel return spring, which absorbs a portion of the recoil energy. The rearward movement of the piston in the buffer cylinder displaces the oil from the rear, to the front of the piston. As the holes in the piston are closed by the pressure of the oil on the valve, the rate of flow of oil is limited by the size of the ports in the edges of the valve and piston. The size of the ports can be varied and the rate of recoil controlled by rotating the oil

buffer cylinder which thus varies the rate of fire of the gun. The action of the buffer supplements the action of the barrel return spring, in bringing the barrel and barrel extension to rest after they have recoiled about 1⅛ in.

68. The first movement of the recoiling portions to the rear causes the transporter, which is engaging a new round, to withdraw the round from the belt and, as the breech block leaves the barrel, the empty cartridge case is extracted from the chamber. During the rearward travel of the breech block, the transporter is supported by the rear transporter cam until it contacts the transporter ramp. The transporter ramp forces the transporter downwards, and the rear of the rear transporter cam is depressed to be returned by its spring when the transporter is clear. The downward movement of the transporter positions the round, which is supported by the claw and ejector, in the cartridge way on the face of the breech block, the empty cartridge case being pushed out through the ejector opening by the ejector. At the same time, the cocking lever, engaging the top plate bracket, withdraws and cocks the firing pin, and the feed lever, riding in its cam groove in the breach block, moves the feed slide transversely in the breech cover. The movement of the feed slide causes the feed pawl to ride over and engage the round which is held by the retaining pawl.

69. The rearward movement of the breech block is arrested when it strikes the buffer in the back plate.

Forward action, fig. 15

70. The breech block rebounds off the buffer and, with the assistance of its return spring, is returned to the forward position. The feed lever is returned by the cam groove in the breech block, causing the feed slide to feed in the belt until the first round is against the cartridge stops and the next round is held by the retaining pawl. The link which was freed by the withdrawal of the previous round is ejected from the feedway as the belt moves into position. During the forward movement of the breech block, the live round is fed into the chamber and, when the transporter meets the front transporter cam, the transporter is raised and the ejector lifted clear of the round. The cocking lever is returned to its normal position by the top plate bracket, the firing pin being retained in the cocked position by the sear.

71. When the lug at the rear of the breech block strikes the horns of the accelerator, the accelerator is rotated forward and is disengaged from the barrel extension shank, and the breech block reaches its forward position in the barrel extension. The barrel return spring reacts and the barrel, barrel extension, and breech block are returned together to the firing position. As the barrel extension moves away from the lock frame, the locking piece pin is cleared from the depressors and the locking piece, riding up the incline on the locking piece cam, enters the locking recess in the breech block. The recoiling portions come to rest when the front of the barrel extension meets the rear face of the trunnion block. The locking piece during the final movement rides along the flat on the top of the locking piece cam which retains it in the locking recess to positively lock the breech block when the gun fires.

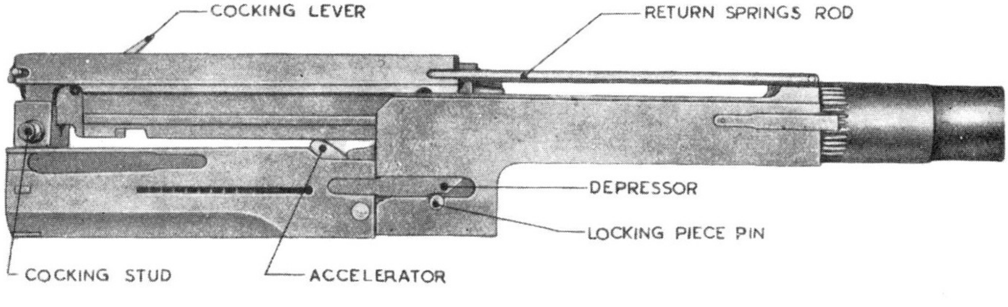

Fig. 15.—Recoiling portions in the rear position

72. During the forward movement of the recoiling portions the oil buffer piston is returned to its normal position by the barrel return spring, and the oil in the buffer cylinder is displaced from the front to the rear of the piston, having free passage through the holes in the piston, the holes having been uncovered by the pressure of the oil on the front of the valve. At the same time, the transporter is lifted upward by the front transporter cam and, at the end of the forward movement,

rebounds off the transporter guide spring in the breech cover, causing the claw and ejector to engage the round which has been positioned by the feed mechanism against the cartridge stops.

73. If the firing control mechanism is operated the sear will be released and the cycle will be repeated.

Safety arrangements

74. The following mechanical devices ensure that the gun cannot be fired before the breech block is locked and the recoiling portions in the firing position.

75. The locking piece cam is arranged to lift the locking piece and lock the breech block to the barrel extension before the recoiling portions reach the firing position.

76. The sear slide cannot be engaged by the trigger motor or electric firing unit until the recoiling portions have reached the firing position.

77. If the sear is accidentally released during the forward travel of the breech block, the cocking lever would take the weight of the firing pin spring and render the force of the blow insufficient to fire the cap.

LOADING AND UNLOADING

Ammunition belt

78. Belts may be issued factory filled or they may be made up locally. When a cartridge is correctly assembled in a link, the distance from the bullet edge of the link to the base of the cartridge case should be between 3·40 in. and 3·45 in. The distance is assured by the setting of the filling or positioning machine, see A.P.1641G. If the box in which factory filled ammunition is issued appears to have been damaged in transit, the belt should not be loaded into an aeroplane ammunition box, but must first be passed through a belt positioning machine.

Loading the gun

79. Ensure that the gun is pointing in a safe direction.

80. The gun must not be loaded until the maintenance operations detailed in A.P.1641L, Vol. II, Part 3, Sect. 2, have been completed.

81. Ensure that the mechanism is correctly assembled for the required direction of feed.

82. When loading the ammunition belt into the ammunition box, ensure that the double loop of the first link will be towards the gun, no matter whether the feed is from the right or from the left.

83. Ensure that the firing control mechanism is set to SAFE and that the breech cover is properly closed and secured by its catch.

84. Push the first round of the ammunition belt past the retaining pawl and ensure that it is held by the pawl. It may be necessary to raise the breech cover to enable this operation to be carried out; if so, care must then be taken to ensure that the breech cover is properly closed and secured.

85. Pull the loading mechanism to the rear as far as possible, then release it.

86. Repeat the operation and the gun is loaded.

87. With synchronized guns, the belt should be brought into the position on the retaining pawl and the pilot should complete the operations 85 and 86 in the air to load the gun.

Note.—Synchronized guns must not be loaded until the engine is firing evenly, as there is a danger of one round being fired from each gun when the engine is started and the shots being early would hole the propeller.

Unloading the gun

88. Ensure that the firing control mechanism is set to SAFE.

89. Disengage the catch and raise the breech cover.

90. Remove the belt from the feedway.

91. Pull the loading mechanism to the rear, hold it in this position and remove the round from the face of the breech block.

92. Release the loading mechanism, ensure that the gun is pointing in a safe direction and fire.

93. The gun is now unloaded.

INSTRUCTIONS FOR USE

94. A fixed gun is fired by setting the firing control mechanism to FIRE, closing any safety switch, or selector switch, which may be incorporated in the circuit, then operating the firing button, or switch, on the control column. A free gun is fired by setting the safety catch on the gun to FIRE and then pressing the trigger.

95. In the event of a misfire, the misfired round can be ejected and the gun released by pulling the loading mechanism to the rear as far as possible and releasing it.

96. A runaway gun may be stopped, if it is within the reach of the firer, by opening the breech cover.

97. Synchronized guns within reach of the pilot must be unloaded before he leaves the cockpit.

BROWNING ·50 in. GUN, No. 2, Mk. II

GENERAL DESCRIPTION

98. The No. 2, Mk. II, ·50 in. Browning gun, see fig. 1, differs from the No. 1, Mk. II, ·50 in. Browning gun in being equipped with a different back plate, in having a mechanically operated trigger

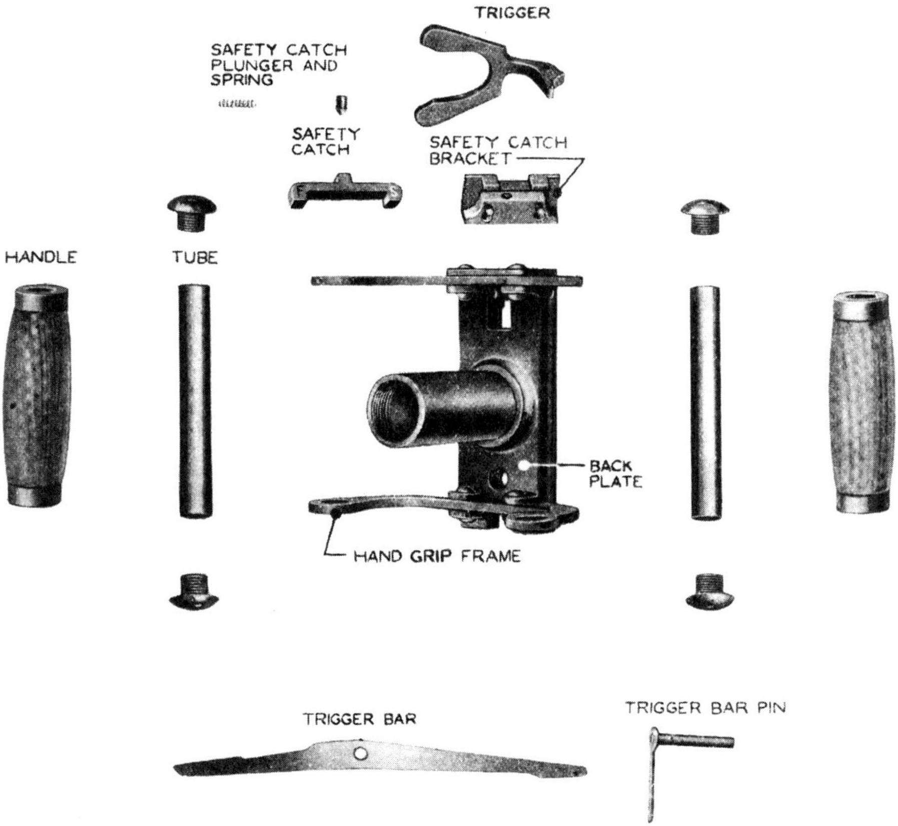

Fig. 16.—Back plate Browning ·50 in. gun, No. 2, Mk. II

mechanism and in the type of loading mechanism employed. These differences are to meet the requirements necessitated by the gun being mounted in a free position. In all other respects the two guns are identical.

Back plate, fig. 16

99. The back plate differs from the back plate of the No. 1, Mk. 2 gun in being provided with a double hand-grip, trigger, and safety catch.

100. The upper and lower frames of the hand-grip are riveted in place of the filler pieces at the top and bottom of the back plate, and the trigger pivots on a pin passing transversely through the back plate. The trigger projects into the rear of the breech casing, is spring-loaded, and engages the trigger bar. The safety catch is mounted on a bracket in the rear of the back plate and has a spring-loaded plunger to retain it in the SAFE or FIRE position. In other respects the back plate is similar to that of the No. 1, Mk. 2 gun.

Trigger bar

101. The trigger bar is mounted on a pin passing through the side plate of the breech casing and the top plate bracket, the pin being secured by a bayoneted joint and leaf spring. The forward end of the trigger bar engages the sear and the rear end is engaged by the trigger.

Loading mechanism

102. The No. 3 loading mechanism equipped with handle, see para. 58 to 60, is used with the No. 2, Mk. 2 gun.

BROWNING, ·50 in. GUN, No. 1, Mk. 1

GENERAL DESCRIPTION

103. The No. 1, Mk. 1, ·50 in. Browning gun, (*fig. 17*), is generally similar to the No. 1, Mk. 2 gun in construction, but differs in some of its details. The chief differences are as follows:—

(i) The muzzle bearing is secured in the barrel casing by crimping.

(ii) The barrel casing screws into the trunnion block and is not provided with a separate support at the rear but screws direct into the trunnion block.

(iii) There is no trunnion block adapter, trunnion block shim, or retaining plunger, the holes for the front mounting bolt being drilled through the trunnion block.

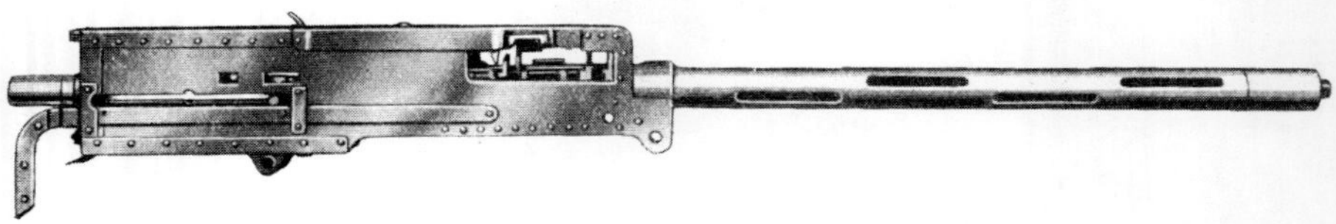

Fig. 17.—Browning ·50 in. gun (M.G.53A), No. 1, Mk. 1, with No. 1 standard loading mechanism

(iv) The barrel has no sleeve, the notches for the barrel locking spring are of square cross-section and the barrel locking spring is shaped to suit the notches.

(v) The back plate buffer tube is screwed into the back plate, being secured by a pin. There is no retainer for the back plate catch.

(vi) The breech cover catch lever is integral with the catch and there is no breech cover detent.

(vii) A separate pivot is provided for the rear transporter cam.

(viii) The breech casing differs in a few details of minor importance.

(ix) The outer end of the cocking stud is flanged.

(x) The No. 1 loading mechanism is used with the Mk. 1 gun.

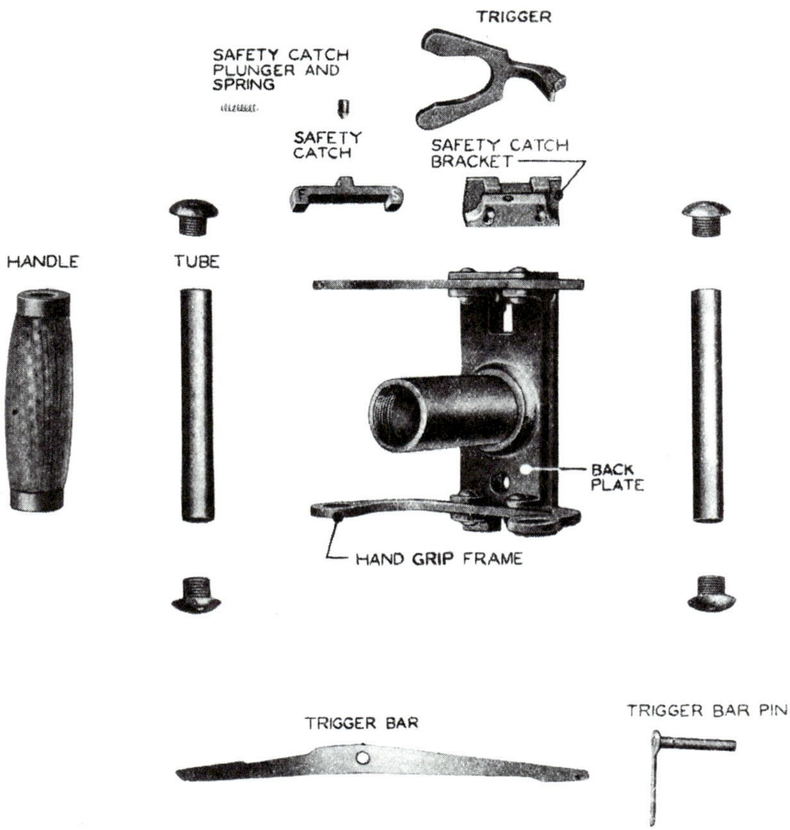

No. 1 loading mechanism

104. The No. 1 loading mechanism consists of a bar which slides in front and rear guide brackets. The guide brackets are fixed to the side plate of the breech casing by screws which are secured by locking wires. The rear guide bracket is handed, a separate bracket being provided for left-hand and right-hand assembly. At the front of the bar is a stop which limits the rearward movement of the bar and near the middle is a hook and eye bolt. The hook engages the cocking stud and the eye bolt is for the attachment of a remote control cable. At the rear of the bar is a handle secured by two rivets, the handle is removed when the gun is remotely controlled.

BROWNING ·50 in. GUN, No. 2, Mk. 1

GENERAL DESCRIPTION

105. The No. 2, Mk. 1, ·50 in. Browning gun differs from the No. 1, Mk. 1, ·50 in. Browning gun in being equipped with a different back plate and in having a mechanically-operated trigger mechanism.

106. The back plate and trigger mechanism are generally similar to those described for the No. 2, Mk. 2, ·50 in. Browning gun, see para. 99 to 101.

INVESTIGATION AND CLEARANCE OF STOPPAGES

General

107. The investigation and clearance of stoppages and the remedies to be applied are divided into two sections as follows, each section being treated separately.

(i) Those stoppages which can be dealt with by the air gunner in the air.

(ii) Those stoppages which can be dealt with only by armament personnel.

108. Stoppages which the air gunner can clear in the air are limited. In a turret he will have two or more guns, and it will be for him to decide, if a stoppage occurs in one gun, whether he will carry on firing with the remaining guns and wait for a suitable lull in the action to clear the gun, which has stopped, or to clear it immediately. If the gun is in a single free gun defence position then it must be cleared immediately. The air gunner must of necessity, learn the action to be taken as a drill in order that, if necessity arises in the heat of battle, he will be able to get his guns firing again as quickly as possible by instinct rather than by having to stop and think what to do. He can only reach and maintain this state of efficiency by constantly practising the drill. The remedies which can be applied by the air gunner are immediate and possibly of a temporary nature only. They will depend on the installation and the spare parts carried in the aircraft.

109. Stoppages which are dealt with by armament personnel on the ground must be treated differently. They must be cleared by a systematic investigation, as the gun may be situated in a remote position in a fighter aircraft, or even in a turret in such a position that the air gunner has had no chance to investigate the stoppage himself. As the armament personnel usually will not know the circumstances in which the gun stopped firing, the quickest and surest method of investigating the cause is by a process of elimination from basic principles. The investigation must be thorough, as it may necessitate reports being made on ammunition or on defective parts of the gun, which may affect design or call for a modification to the part. Remedies applied by the armament personnel may necessitate complete dismantling of the gun, followed by cleaning and gauging to ensure that there is no possibility of the stoppage recurring.

110. Both the air gunner and the armament personnel must bear in mind that stoppage tables are to be used as a guide only. They are completed from experience gained over a long period and can be relied upon, but there are exceptions for which no tables can be laid down, especially if a part breaks and falls amongst the gun mechanism. Both the air gunner and the armament personnel should be alive to this fact when applying the immediate action, or when carrying out the investigation.

111. Stoppages can be reduced to a minimum and should be rare if meticulous servicing and inspections are carried out with the utmost care when they become due. It is essential from this aspect that the air gunner is conversant with the armament personnel's duties and responsibilities, and vice versa.

112. A table of immediate action to be taken by the air gunner will be issued later.

Investigation to be completed by armament personnel

113. Ensure that the firing button is set at "Safe" and if possible disconnect the lead of the firing unit, at the battery.

114. Using a piece of chalk, mark the position of the cocking stud on the side of the gun body. This will enable the stoppage to be reconstructed afterwards, if it is necessary to remove the gun from the aircraft for purposes of investigation.

115. Look for anything abnormal externally, especially as regards the feed of the gun, lay of the belt, alignment of chutes, condition of ammunition boxes, etc.

116. Hold the cocking stud securely so that the stoppage is not cleared and raise the breech cover. Look for anything abnormal internally, such as the position of rounds in the feedway, condition of transporter, feed slide and feed lever.

117. If the cause is apparent through these investigations, the gun can be unloaded and the fault remedied.

118. If the cause is not immediately apparent through these investigations, pull the cocking stud slowly to the rear and catch the round or empty case extracted from the chamber or removed from the face of the breech block.

119. Inspect the gun and ammunition thoroughly to find the cause of the stoppage. It may be caused either by the gun or the ammunition.

Note.—When the stoppage is caused through defective ammunition, the particulars on the base of the case and if possible those on the ammunition box, should be recorded. This will enable an investigation to be made in connection with the ammunition, if the stoppage is repeated.

120. When pulling the cocking stud to the rear, one of three things will happen.

(i) A live round will be extracted.

(ii) An empty case will be extracted.

(iii) Nothing will be extracted.

If a live round is extracted

121. Examine the cap of the cartridge to see whether it has been struck or not, having previously noted the position of the breech block.

122. *If the cap has been fully struck*, a faulty round of ammunition is indicated. Feed must have taken place and so there is no mechanical defect in the gun.

123. *If the cap has been lightly struck*, there must be a defect in the parts of the gun mechanism concerned with the firing of the round. Feed must have taken place so that the fault will be limited to the following:—

Fault	*Remedy*
(i) Insufficient protrusion of the firing pin.	(i) Gauge the protrusion of the firing pin, exchange as necessary, see A.P.1641L, Vol. 2, Part 3, Sect. 2, Chap. 2, para. F64.
(ii) Weak firing pin spring.	(ii) Remove the firing pin spring, see A.P.1641L· Vol. 2, Part 3, Sect. 1, Chap. 2, para. 27, Examine it for fractures and check its free length, see A.P.1641L, Vol. 2, Part 3, Sect. 2, Chap. 2, para. F61 to F63.
(iii) Cushioning of the firing pin.	(iii) Examine the firing pin and firing pin extension for fluting, see A.P.1641L, Vol. 2, Part 1, Leaflet B.2—W.
(iv) Burring of the firing pin hole, or loose firing pin hole bush.	(iv) Clean off burrs, taking care not to enlarge the hole. If firing pin hole bush is loose, exchange the breech block.
(v) Dirt, brass chips, fouling or congealed lubricant in the firing pin hole.	(v) Clean the firing pin hole and the firing pin way in the breech block. If brass chips are present, look for sharp edges on the face of the breech block and cartridge rim guides, likely to tear the cartridge case. Stone down as necessary, see A.P.1641L, Vol. 2, Part 3, Sect. 3, Chap. 2, para. F43.

124. *If the cap has not been struck but the breech block is fully forward,* one of the parts concerned with the firing of the gun must be broken. Feed must have taken place, and as there is no mark on the cap of the cartridge case any of the following faults may have occurred:—

Fault	*Remedy*
(i) Broken firing pin.	(i) Exchange the firing pin complete.
(ii) Broken firing pin spring.	(ii) Exchange the firing pin spring, see A.P. 1641L, Vol. 2, Part 3, Sect. 1, Chap. 2, para. 27.
(iii) Broken or damaged sear, firing pin bent or sear slide.	(iii) Examine the sear for fractures and test it, see A.P.1641L, Vol. 2, Part 3, Sect. 2, Chap. 2, para. F68 to F70. Examine the firing pin bent for fractures, see A.P.1641L, Vol. 2, Part 3, Sect. 2, Chap. 2, para. F58. Examine the sear slide and ensure that it operates the sear to release the firing pin, see A.P.1641L, Vol. 2, Part 3, Sect. 2, Chap. 2, para. F71 to F73. Exchange components as necessary and test, see A.P.1641L, Vol. 2, Part 3, Sect. 2, Chap. 2, para. F114 and F115.
(iv) Failure of the electric firing unit.	(iv) Test the adjustment of the stroke of the trigger of the firing unit concerned, see A.P.1641N, Vol. 2, Part 3, Sect. 1, Chap. 4. If the above test is not satisfactory carry out the electrical test, see A.P.1641N, Vol. 2, Part 3, Sect. 1, Chap. 4, para. 50 to 55.

125. *If the cap has not been struck and the breech block is held to the rear,* a mechanical defect is preventing the moving parts from going fully forward. Feed must have taken place and the fault may be due to any of the following:—

Fault	*Remedy*
(i) Incorrect adjustment of cartridge head space. The gun has been overbreeched.	(i) Adjust the cartridge head space, see A.P.1641L, Vol. 2, Part 3, Sect. 1, Chap. 2, para. 81 to 98.
(ii) Burrs in the front barrel bearing or the barrel bearing in the trunnion block.	(ii) Remove all burrs from both bearings, and rust if present from the rear bearing, see A.P.1641L, Vol. 2, Part 3, Sect. 2, Chap. 2, para. F94 and F103.
(iii) Broken or weak return spring.	(iii) If either of the return springs are broken they must be exchanged. If the springs appear serviceable, test their weights against new springs.
(iv) Defective return spring rod.	(iv) If the return spring rod is bent or broken, the rod and springs must be exchanged complete. No attempt must be made to straighten a bent return spring rod.
(v) Burrs on any part of the recoiling portions of the gun.	(v) Inspect the recoiling portions for burrs and stone as necessary. Pay particular attention to the barrel bearings, locking piece, and locking piece cam, and the breech block guides. Test each part for freedom of movement in its corresponding component part.
(vi) Broken or weak barrel return spring.	(vi) If the barrel return spring is broken it must be exchanged. If the spring appears to be serviceable, test the weight against a new spring.

	Fault		*Remedy*
(vii)	Separated case, usually determined by the difficulty experienced in pulling the recoiling parts to the rear when cocking the gun, and may be due to:—	(vii)	Remove the separated case using the clearing plug.
	(*a*) Incorrect adjustment of cartridge head space. The gun has been underbreeched.		(*a*) Adjust the cartridge head space, see A.P.1641L, Vol. 2, Part 3, Sect. 1, Chap. 2, para. 81 to 98.
	(*b*) Defective cartridge case. Flaw in the metal or incorrect hardness of the case.		(*b*) Nil.
(viii)	Mal-adjustment of the oil buffer setting. The buffer has been set at less than three notches from "O" or "C".	(viii)	Adjust the speed regulator, see A.P.1641L Vol. 2, Part 3, Sect. 1, Chap. 2, para. 78 to 80.
(ix)	Oil buffer detent loose, allowing the accelerator to chatter.	(ix)	Examine the oil buffer detent and ensure that it is a tight fit, see A.P.1641L, Vol. 2, Part 3, Sect. 2, Chap. 2, para. F27. Exchange if necessary.

If an empty case is extracted

126. This indicates that either extraction or ejection have failed after the last round was fired. Feed must have taken place and firing was satisfactory, therefore one of the following faults may have occurred.

Note.—Inspect the case for any marks likely to give an indication of the stoppage.

	Fault		*Remedy*
(i)	Hard extraction of the case caused by either a dirty chamber or a defective case. *Note.*—A part of the rim of the case may be torn off.	(i)	If the case is difficult to remove, the cleaning rod should be used to push the case out. Clean the chamber thoroughly, and inspect it for burrs or damage; exchange the barrel as necessary.
(ii)	Defective ejector. This should only occur for the last empty case.	(ii)	Exchange the ejector.
(iii)	Weak charge.	(iii)	Obtain details of the ammunition and box and report.

If nothing is extracted

127. The fault must be in the loading operation of the gun and may be any of the following:—

	Fault		*Remedy*
(i)	Incorrect timing of the firing unit. The sear is released before the recoiling parts are fully forward and the claw of the transporter could not grip the base of the next round in the belt.	(i)	Adjust the timing of the sear so that it is released correctly, see A.P.1641N, Vol. 2, Part 3, Sect. 1, Chap. 4, para. 20 to 28 or para. 41 to 49 according to the type of firing unit used.
(ii)	Insufficient clearance in the feedway.	(ii)	Check the clearance in the feedway with the ·868 in. feedway gauge, see A.P.1641L, Vol. 2, Part 1, Leaflet B.3—W.
(iii)	Badly filled belt.	(iii)	Check the adjustments of the belt positioning machine, see A.P.1641G, Vol. 2, Part 3, Sect. 1, Chap. 6, para. 54 to 58.
(iv)	Link jammed in the feedway.	(iv)	Inspect the link chutes for distortion, or misalignment, and adjust or exchange as necessary. Empty the link chutes and tanks if full.
(v)	Defective feed pawl spring.	(v)	Examine the spring for fractures, check its free length against a new spring and exchange as necessary.
(vi)	Defective retaining pawl spring.	(vi)	Examine the spring for fractures, check its free length against a new spring and exchange as necessary.
(vii)	Defective transporter claw. *Note.*—Due to the action of the feed pawl leg, the breech block will stop in the firing position.	(vii)	Exchange the transporter.
(viii)	Defective feed lever.	(viii)	Exchange the feed lever.

Amendments

These amendments, issued subsequent to the initial printing of the original handbooks, were intended to be pasted by the user on to the pages in question.

Amendment to Handbook on the Vickers Gun
Page 18, add to end of para (c):
"All crank handles are now hardened and tested prior to issue. A certain amount of hammering up of this part must be expected, but the actual amount to be done can usually be kept within safe limits by careful adjustment and tuning up of the gun. A slight increase of the tension on the fusee spring will usually reduce the hammering but if in spite of precautions the crank handle does become flattened and burred, the burr should be removed by careful filing, including a slight radiusing of the top edges. The steel of which crank handles are made is not a case-hardening steel, and should not therefore be treated locally. If, after the exercise of the above precautions, the trouble still persists in a marked degree, it may be necessary to return all the existing crank handles and issue new case-hardened ones in their place."

Amendments to Browning .303in Mark II Gun
Page 86, para 9, add to end of first sentence:
"or muzzle attachment choke".
Page 87, add para 10A:
"Muzzle attachment choke
10A. The muzzle attachment choke is a screwed plug and is the same shape as the small end of the flash eliminator. If the flash eliminator is not required the muzzle attachment choke must be used and the same care must be exercised when fitting and cleaning."
Page 100, add para 69A:
"69A. When belts are issued factory filled in Boxes A.S.A. H.13, the following precautions must be observed. If the box appears to have been broken in transit, the belts therein should not be directly loaded into an aeroplane ammunition box, but should first be passed through a belt positioning machine."

Amendment to Hispano 20mm Mk I and II Guns
Page 162, insert at end of para 121:
Note.—This does not preclude the use of the recocking lever by the gunner, when one is carried to attend to the guns."